° 5 ⌐

IMAGES
of Aviation

HUNTSVILLE
AIR AND SPACE

D1262845

This historical marker located in the community of Sulphur Springs in Madison County marks the site of the Quick airplane construction and flight, the first to be flown in the state of Alabama. It was among the first monoplanes to be flown in the United States when it went airborne in April 1908. (Author's collection.)

ON THE COVER: Curtiss Quick, Huntsville native and member of Huntsville's Quick family of early southern aviationists, bought an OX-5 powered Canuck (Canadian built JN4D "Jenny") in 1920 in Pulaski, Tennessee, and at age 26 taught himself and his brothers to fly in a field near their home outside Huntsville. Seated is Cady Quick, his sister, who became a pilot at age 23, and is one of Alabama's first and youngest female pilots. Curtiss was a test pilot, exhibitionist, aerial photographer, and crop duster, and operated his Quick Duster's Flying Corporation throughout the South. (Rick Clark.)

IMAGES
of Aviation

HUNTSVILLE
AIR AND SPACE

T. Gary Wicks

ARCADIA
PUBLISHING

Copyright © 2010 by T. Gary Wicks
ISBN 978-0-7385-6607-8

Published by Arcadia Publishing
Charleston SC, Chicago IL, Portsmouth NH, San Francisco CA

Printed in the United States of America

Library of Congress Control Number: 2009924094

For all general information contact Arcadia Publishing at:
Telephone 843-853-2070
Fax 843-853-0044
E-mail sales@arcadiapublishing.com
For customer service and orders:
Toll-Free 1-888-313-2665

Visit us on the Internet at www.arcadiapublishing.com

To engineering, and the men and women who have used their innate skills and acquired knowledge and applied themselves, with discipline and dedication, to the design and construction of flying machines and space vehicles, which have carried humans not only to all parts of the globe, but beyond earth's atmosphere—to walk on the surface of the moon.

CONTENTS

ACKNOWLEDGMENTS

The author gratefully acknowledges the support, encouragement, and resources provided by the following individuals and organizations in making this book possible: Billy Singleton and Wayne Novy, Southern Museum of Flight; Alan Renga, San Diego Air and Space Museum (SDAM); Bill Yenne, aviation author; Public Affairs office, NASA Marshall Space Flight Center; Marshall Retirees Association; Experimental Aircraft Association, Chapter 190 members including Jules Bernard, Rob Maulsby, Ann Maulsby, and Harold McMurran; Irene Willhite, archivist, U.S. Space and Rocket Center (USS and RC); Anne Coleman, archivist, University of Alabama in Huntsville (UAH) Library; University of Alabama Map Library, Tuscaloosa, Alabama; Richard Tucker, director and Twearnier Rice Huntsville Airport Authority; Rick Davis, director, Cummings Research Park; Roger Jones, Madison County commissioner, District 1; Ray Jones, G. W. Jones and Son Consulting Engineers; Ray Watson, retired, Teledyne Brown Engineering (TBE); Joe Thurman, retired, SRS Technologies; Steve Doyle and Dave Dooling, *Huntsville Times* writers; Rhonda Larkin and Donna Dunham, Madison County Records Center; Paul Freeman, airfield researcher; Mike Sparkman, Johnny Johnston and Tony Campbell, retired, Huntsville airport; Gerry Moore; James B. Hill Jr.; Carole Record; and Quick family descendents: Evelyn Clark, Rick Clark, Valera McGuire, Ron Wicks, Kathy Sparks, Bob Quick Jr., Frances Case and Marilyn Snowden.

The author is especially grateful and deeply indebted to historian Michael Baker and Claus Martel of the U.S. Army Aviation and Missile Command (AMCOM), Redstone Arsenal; historian Mike Wright and Roena Love of the NASA Marshall Space Flight Center (MSFC); archivist Ranee Pruitt, Thomas Hutchens, and staff of the Heritage Room and Archives of the Huntsville-Madison County Public Library; and Bob Adams and Nancy Rohr, Huntsville Madison County Historical Society. Their guidance, encouragement, and contributions helped to make this project a reality.

And lastly, I thank my wife, Peggy, and son, Gabe, without whose support and assistance I would not have undertaken the challenge this book presented.

INTRODUCTION

The evolution of air and space in Huntsville, Alabama, spans 100 years, from early experimentation with a piloted flying machine at the beginning of the 20th century, to rocket development that launched America's first astronauts in the middle of the 20th century and put them on the moon. This evolution continues with a rocket plane that delivered the space station and crews to permanently orbit the earth as we entered the 21st century. Before there were rockets that boosted men into space, there were airplanes that lifted them into the air. The manned rocketry and spacecraft of the 1950s were built on the 50 prior years of airplane technology development. In 1903, the Wright brothers were the first in the world to become airborne. While but for a short distance into the ocean winds of Kill Devil Hills, North Carolina, it was the first controlled and sustained flight of a powered, manned craft. In 1926, Robert Goddard, a U.S. physics professor, launched the world's first liquid-fueled rocket in Auburn, Massachusetts.

The state of Alabama's entry into the air age began in Madison County—a homegrown product—largely due to the imagination, effort, and influence of one man, William Quick. A contemporary with early aviation pioneers, this Madison County inventor started work on a flying machine before 1900. While early experimentation by Quick with a 1908 prototype demonstrator was modestly successful, he filed for a patent for an improved flying machine in 1912 for the purpose of forming an aircraft production company. He was a man obsessed with flying, and this was the beginning of the Quick family's infatuation with airplanes. All of his seven sons and one daughter were aviation enthusiasts and all eight became licensed pilots, accomplishing many feats in the field of aviation. Three sons had numerous patents in connection with aviation. Ann Quick, their mother, also learned to fly. Another son designed and built one of the first all-metal airplanes in the country. Terah Maroney, Ann's brother, was familiar with his brother-in-law's experiments and became interested in powered flight. Maroney moved to Montana in 1911, where he later started a one-plane airline, flying a Curtiss-designed plane. One of his passengers in Seattle on July 4, 1914, was a 24-year-old lumberman named William Boeing. It was said that Boeing was so taken by that first airplane ride that he learned to fly himself during the next two years. He formed a company; built and flew the B and W, the first Boeing airplane; and incorporated the Boeing Company on July 15, 1916.

The Quick family's early developments and their barnstorming exhibitions during the 1920s comprised most of Madison County aviation activity in the first quarter of the 20th century. However, the beginning of the 1930s saw some movement in the city of Huntsville. In 1931, dreams came true for local aviators led by Thomas Quick and the Jaycees. The city cleaned off a field and converted it into Huntsville's new 150-acre airfield, located west of Alabama Street between Bob Wallace and Thornton Avenues. The airfield was dedicated by Lorraine Quick, Miss Junior Chamber of Commerce, by dropping a bottle of wine on the field from a Stinson cabin plane. An unexpected boost to the airfield came in 1938, when a squadron of 18 Seversky P-35 army pursuit planes out of Selfridge Field in Michigan ran into bad weather north of Birmingham while en route to maneuvers in Florida. The unit turned around and headed back toward Huntsville, where residents got in their cars and headed out to Whitesburg Drive where their headlights illuminated the grass strip. All of the army craft came down safely just before the weather closed in.

The Huntsville-Madison County Airport got its start in 1940 when the city and county jointly purchased some 400 acres of land south of Huntsville for the construction of a new airport to replace the airfield off Whitesburg Drive. The phased development of the new facility was started in 1941 as a Works Progress Administration (WPA) project to pave two runways. On November 5, 1944, which marked the official opening of the Huntsville-Madison County Airport, the *Huntsville Times* ran a full-page advertisement from Pennsylvania Central Airlines (PCA) proclaiming that the "air minded" city was now on the "main line". The DC-3 *City of Huntsville* arrived from Birmingham and Susie Spragins christened the airplane with a bottle of water from Huntsville's Big Spring. In 1948, steps were undertaken by the city and county to widen and extend the runways. In 1952, the administration building was completed. The north-south runway was extended in 1955 to enable the use of the airport by four engine commercial airliners.

The 1950s saw the beginning of growth ignited by the influx of the German rocket team led by Dr. Wernher von Braun. The escalating cold war with Russia saw the army use Redstone Arsenal to work on missiles and missile defense, but it was in the 1960s that a small town, known for cotton and watercress, became a leader in the field of rocket technology. Huntsville would come to be known as the Rocket City, as it led the United States in its quest for space. Thousands of engineers, research scientists, and technicians would come to work in Huntsville, helping make the city the Space Capital of the Universe. From 1950 to 1958, Huntsville's population grew from 16,000 to 150,000, and Madison County grew from 70,000 to 200,000. As the army's Missile Command (MICOM) and NASA's Marshall Space Flight Center (MSFC) expanded, they attracted numerous major aerospace companies and dozens of supporting industries. New residential housing developments and more schools, utilities, and roads increased the demand for city and county resources. The airport authority took over control of the Huntsville Madison County Airport in 1956. With an inadequate, un-expandable airport and no money, the authority sold the city airport and, with added state and federal funds, bought land and built runways and support facilities for a new airport near the city of Madison. The new Huntsville-Madison County Airport opened on October 29, 1967. Originally called the Jetport or Jetplex, the name changed to the Huntsville International Airport in April 1988.

The outbreak of World War II in Europe in 1939 and escalating global tensions were events that would shape Huntsville's future. Rep. John Sparkman of Huntsville was influential in getting the U.S. Army Chemical Munitions Service to locate a $40-million plant in Madison County—the Huntsville Arsenal; and a second facility shortly after—the Redstone Ordnance Plant for the U.S. Army Ordnance Corps. At the end of the war, the need for the two plants also ended. But in 1948, Sparkman, now United States senator, again used his influence and the two were combined and reactivated as Redstone Arsenal for Ordnance Corps research on guided missiles. A team of army and German rocketeers from Fort Bliss, Texas, was transferred to Huntsville. They had built the V-2, a vengeance weapon for Hitler's *Wehrmacht*. Developments such as the Hermes, Corporal, Honest John, and Nike missile systems provided the U.S. Army with antiaircraft defense weaponry and heavy ordnance delivery capability, including nuclear payloads. Increased distance and payload were provided by the evolving larger Redstone and Jupiter delivery systems, under development by the U.S. Army Ballistic Missile Agency (ABMA) formed in 1956 under the command of Maj. Gen. John Medaris. On October 4, 1957, the Soviet Union placed an artificial moon, *Sputnik 1*, in orbit. But on January 30, 1958, the U.S. Army's Jupiter C—a modified Redstone missile outfitted with upper stages—placed a small satellite, *Explorer 1*, built at the Jet Propulsion Laboratory in Pasadena California, into orbit.

The U.S Army's ABMA became the Missile Command (MICOM) in August 1962. In addition to the Redstone and Jupiter, the 1950s and 1960s saw the army develop four basic classes of missiles: battlefield, antiaircraft, antitank, and antimissile. Following Redstone and Jupiter there was Pershing, a solid propellant, nuclear-tipped missile that became important to the Western allies in the European theater. Others included Multiple Launch Rocket System (MLRS) and Army Tactical Missile System (TACMS). To protect against enemy aircraft there were Nike Ajax, Nike Hercules, Hawk, Chaparral, Redeye, and Stinger, followed in the 1980s with the advanced Patriot. MICOM developed a variety of antitank weapons, including TOW and Hellfire. A missile capability to destroy other missiles grew out of the antiaircraft programs. In June 1960, a Nike Hercules guided missile tracked and shot down a Corporal ballistic missile, the first time a missile was killed by a missile. Earlier in the year, a Hawk missile had shot down an Honest John, a short-range unguided missile. In 1962, a Nike Zeus and a Titan II etched the famous X in the sky over Kwajalein Atoll in the South Pacific and proved that a missile could intercept an Intercontinental Ballistic Missile (ICBM). While Safeguard was initially designed as a nuclear defense umbrella for the entire nation, the U.S. Army Ballistic Missile Defense organization continued this research to field more advanced systems in the 1980s, including the landmark impact kill of a Minuteman I target reentry vehicle over the Pacific Ocean by the Homing Overlay Experiment in June 1984. The mission of Redstone Arsenal was dramatically expanded with the army's decision to relocate the Aviation Troop Command from St. Louis, Missouri, to Redstone Arsenal. The resulting Aviation and Missile Command (AMCOM) was formally established on October 1, 1997, co-locating the management of army aviation platforms and their armament missiles at Redstone Arsenal.

The history of aviation at Redstone Arsenal can be traced to the installation's World War II beginnings. Originally part of the Chemical Corps' Huntsville Arsenal, the Redstone Army Airfield was established to accommodate the planes used to test incendiary munitions manufactured during the war. With the cessation of hostilities in 1945, the Redstone Army airfield began an 11-year hiatus during which time no planes were assigned to it. In July 1956, ceremonies marking the opening of the renovated and expanded Redstone Army Airfield were held, dedicating the airfield and newly opened control tower. The airfield's operation has supported the aviation and missile research and development efforts at Redstone Arsenal over the years and shares the field with MSFC for its civilian space activities. Four U.S. presidents have landed in *Air Force One* at Redstone Army Airfield.

MSFC was activated on July 1, 1960, with the transfer of buildings, land, space projects, property, and personnel from the U.S. Army, and Wernher von Braun became the center's first director. The center was named in honor of Gen. George Marshall, the U.S. Army Chief of Staff during World War II. In 1961, MSFC's Mercury-Redstone vehicle boosted America's first astronaut, Alan Shepard, on a suborbital flight. MSFC's first major program was development of the Saturn rockets. The first Saturn I flew in 1961; it and nine others, followed by nine more powerful Saturn IB rockets, worked perfectly. The first Saturn V, equipped with live second and third stages and an Apollo spacecraft rebuilt after the tragic on-pad 1966 fire, was launched on a maiden voyage on November 9, 1967, followed on December 21, 1968, by *Apollo 8* crew to lunar orbit and back. On July 16 1969, the sixth Saturn V, carrying *Apollo 11*, left earth for the moon. Four days later on July 20, 1969, Neil Armstrong stepped off the lunar module onto the Sea of Tranquility—the first human to set foot on the moon. MSFC also developed the Lunar Roving Vehicle for transporting astronauts on the lunar surface on the last three Apollo lunar missions.

In 1973, MSFC's Saturn V rocket lifted Skylab, the first crewed orbiting space station. As part of Skylab, Marshall had responsibility for many scientific experiments, the development of the Skylab Orbital Workshop, the Apollo Telescope Mount, and the Skylab Multiple Docking Adapter. In 1975, a Saturn IB rocket lifted the Apollo spacecraft into Earth's orbit for the historic linkup with the Russian-Soyuz spacecraft. That mission also included experiments provided by MSFC scientists. Three High Energy Astronomy Observatories to study stars and star-like objects were launched in 1977, 1978, and 1979.

In the early 1970s, MSFC was assigned responsibility for developing the space shuttle propulsion elements, including the external tank, solid rocket boosters, and the space shuttle main engines. In 1978, the space shuttle orbiter "Enterprise" arrived at MSFC for vibration testing. In 1983, NASA launched its first Spacelab mission onboard the space shuttle, a scientific lab managed by MSFC and carried in the payload bay of the orbiter. More than 20 Spacelab missions were conducted over the next 15 years. In 1990, the MSFC–developed Hubble Space Telescope was launched. In 1998, NASA launched the first U.S. Space Station element—the Unity node, built by the Boeing Company at MSFC.

Huntsville Research Park began in 1961 with von Braun's idea that a research institute in Huntsville would create an atmosphere attractive to professional and scientific groups. He also felt that an industrial park should be developed adjacent to this institute to provide a place where companies could locate and fulfill the research and development requirements of MSFC and the Missile Command. Von Braun's idea soon became a reality when the City of Huntsville zoned approximately 2,000 acres as a Research Park District and prohibited any development in this area other than industrial.

In the late 1960s, von Braun approached the Alabama legislature with the idea of establishing, with the help of the Missile Command and MSFC, a museum that would provide a showcase for the hardware of the space program. In 1968, state lawmakers and citizens voted to finance this "showplace of America's space program." The Space and Rocket Center's rocket park contains the world's most comprehensive collection of military and space rocketry. There is a vertical Saturn I, as well as an imposing 363-foot Saturn V. The Davidson Center for Exploration opened in 2008, housing a completely renovated Saturn V moon rocket and exhibits of future manned space exploration vehicles, which will replace the space shuttle. The history of air and space in Huntsville can be traced from its earliest beginnings in aviation through world-shaking achievements in rocketry, missilery, airborne, and space systems. Its centerpiece is the Saturn V vehicle built by the von Braun team of rocket scientists, which put the first men on the moon, arguably the world's greatest technological achievement—ever.

One

QUICK FLYING MACHINE

The Quick Flying Machine is a mid-wing monoplane with an upright pilot position, fuselage mounted engine, direct drive propeller, and three-wheel landing gear. Unique features include take-off under its own power, wing warp, and pitch control. The original monoplane, restored by the Experimental Aircraft Association, is shown here on display in the Davidson Center for Exploration at the U.S. Space and Rocket Center in Huntsville. (Author's collection.)

Quick began building a flying machine, an idea coming forth from years of preoccupation with inventions. An illustration of the old shop and forge where the plane was fabricated and assembled shows the plane under construction by Quick and his three sons. The shop was used for woodworking, especially furniture making, and the attached forge for metalwork—farm implements, horseshoes, and wagons. (Author's collection.)

The shop and forge were located upstream of the dam. The Quick family dug a millrace and diverted part of the Flint River's flow to a gristmill where the water's 10-foot head drove a turbine and provided mechanical power to the mill and the shop. Ann Quick and her son Joe are shown standing on the dam with the shop in the background. (Evelyn Clark.)

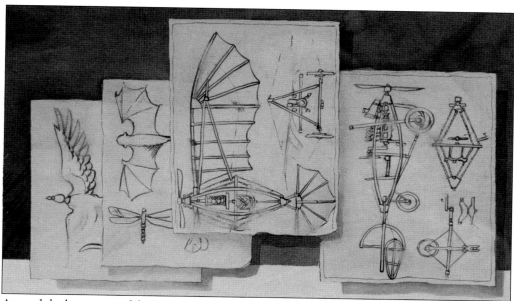

Around the beginning of the 20th century, Quick developed a serious interest in aerial "navigation" as he called it, and designed a flying machine that certainly was in harmony with nature's system for accomplishing the feat. Coupled with his self-study of biology, mechanics, and physics, he fashioned his plane after the bats, buzzards, and dragonflies, which comprised his world of naturalist thought. Shown is his concept evolution. (Author's collection.)

Handwritten in pencil on small notebook paper to Munn and Company, Broadway, New York, and tucked away within the pages of one book in Quick's technical library entitled "Practical Pointers for Patentees," Quick's 1903 letter, shown here, requests a one year subscription of *Scientific American*, a copy of the book *Practical Pointers for Patentees*, and a copy of the book *Experiments in Aerodynamics* by S. P. Langley. (Author's collection.)

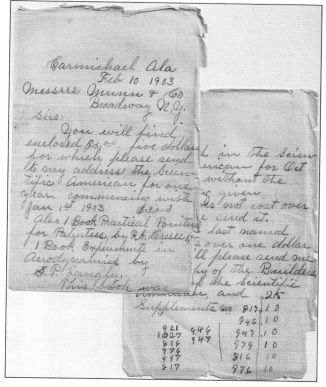

Upon completion, Quick's 16-year-old son William took off down the field in the plane's flight test as illustrated. It bounced once or twice, and then it lifted off the ground. The young pilot became excited, leaned forward, and looked down to see how high he was from the ground. As William leaned forward, the plane went down as designed. "The ground came back up too quick," he later said. (Author's collection.)

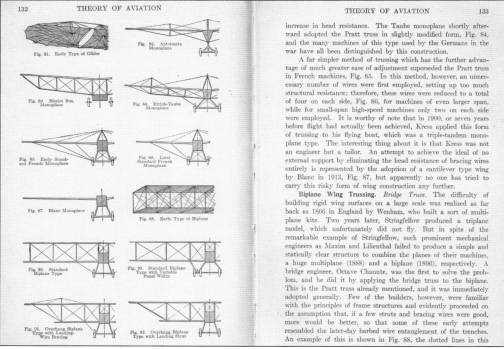

This book from Quick's library entitled *Practical Aviation, Second Edition*, Hayward, 1919, covers the current aviation knowledge at that time, including the theory of aviation, the power of propellers, aviation motors, and types of early airplanes—both biplanes and monoplanes. Also included in the book is the construction of airplanes, like Wright, Curtiss, and Bleriot designs, and the evolution of wing trussing up to that time. (Author's collection.)

In its demonstration flight, the machine tipped down sideways, landed at a slant, and the wheels were torn off. The plane was hauled back to the shop, but that was not the end of inventions for Quick. Being the inventor that he was, he went to work on an improved flying machine, receiving a patent, shown here, on October 21, 1913, for the purpose of manufacture and sale. (Kathy Sparks.)

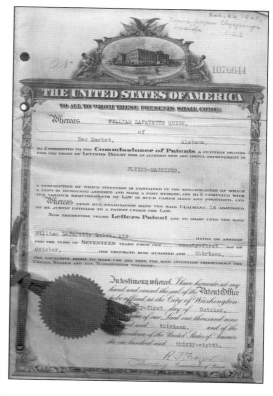

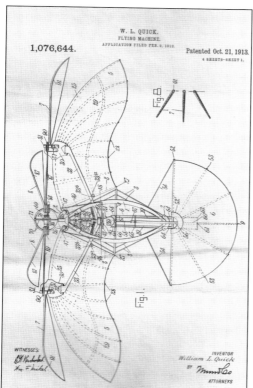

Shown are two of the patent's nine figures providing for, according to the patent, "a full, clear, and exact description." Detailed are the machine's design features—retractable landing gear, folding wings, ornithopter method of thrust propulsion, and its operational features—"the constituent parts are so arranged as to give automatic stability to the entire apparatus, while at the same time permitting any manual adjustments which may be required." (Kathy Sparks.)

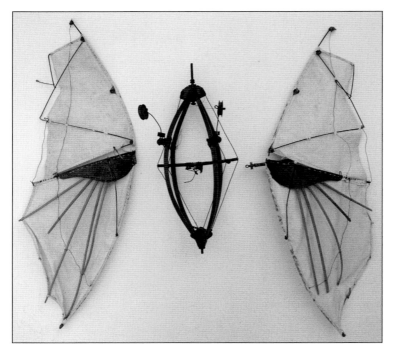

A scale model was constructed by Quick for the purpose of visualizing and explaining his patented invention. Shown with wings detached to display their ribbed construction and with fuselage inverted to show landing gear, his invention, according to the patent, "relates to flying machines, and it comprises a new and improved construction of this sort in which great strength and lightness are combined with perfect elasticity." (Rick Clark.)

Y MERO

VEDNESDAY, JUNE 26, 1912.

NEW KIND OF MONOPLANE

Wm. Quick Invents Bird Like Machine

CLAIMS IT WILL BE IMPOSSIBLE FOR IT TO FALL DIRECTLY TO THE GROUND

William L. Quick of this county has invented and perfected a new type of monoplane which, it is believed, will

Quick's commercialization interests appear in this article in the Wednesday June 26, 1912, *Weekly Mercury* entitled "NEW KIND OF MONOPLANE, Wm. Quick Invents Bird Like Machine" where it is noted that "he has twenty patents pending, sixteen of which have already been granted" and how he "expects to organize a company for the manufacture of his machine and he would like to have the plant located in Huntsville." (Madison County Records Center.)

COPYRIGHT BY O.E. GRAVES

Terah Maroney, Quick's young brother-in-law and Huntsville native, helped Quick build his flying machine, and through this experience, Maroney built one of his own planes in Great Falls, Montana. He made his first flight there on July 6, 1911—the first man from the state of Montana to fly. Maroney attended the State Fair in Helena, Montana, in 1911, where he assisted Cromwell Dixon in his exhibition flights. (SDAM.)

Maroney was able to raise sufficient capital to enroll in a flight-training course in the Curtiss School in North Island, San Diego, California. He purchased a used Curtiss airplane and was issued an FAI license on March 14, 1912. As recounted in the book, *Boeing Planemaker to the World*, Maroney took William Boeing, founder of the Boeing Company, on his first plane ride, July 4, 1914. (Bill Yenne.)

BOEING
PLANEMAKER TO THE WORLD
REVISED AND UPDATED

REDDING & YENNE

On 4 July 1914, a barnstormer named Terah Maroney was hired to put on a flying exhibition as part of the Independence Day festivites in Seattle, Washington. Having put on a display of aerobatics in his Curtiss seaplane, Maroney landed and offered to take up passengers. Almost on impulse a lumber company owner name Bill Boeing stepped up and allowed as how he'd like to take a ride.

What happened in the next 30 minutes changed the course of aviation history and the history of the Northwestern United States as well. William E Boeing had caught the flying bug. He was never to recover.

William and Ann Quick had seven sons and three daughters, shown in this early 1900s photograph. All of the seven sons and one daughter were aviation enthusiasts and all eight became licensed pilots. Ann also learned to fly. This was the beginning of the Quick family's infatuation for airplanes. They were truly a pioneer flying family. (Frances Case.)

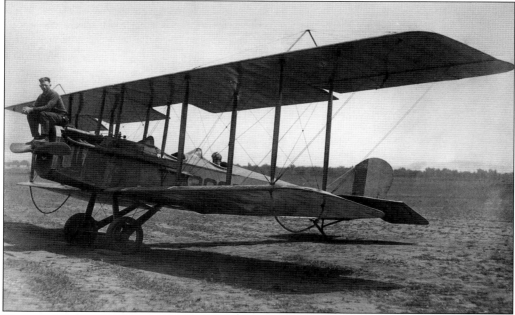

Thomas Quick, an early test pilot, is shown sitting atop a Hisso Standard biplane (Standard J-1 re-engined with an Hispano-Suiza engine) purchased in Huntsville on December 13, 1922, as war surplus. This photograph was taken just off present-day U.S. Route 231 North at Walker Lane. Thomas was the second Quick boy to learn to fly and taught his other brothers to fly, including the eldest, Carrol, and the youngest, Spencer. (Rick Clark.)

In 1921, brothers Erie, Curtiss, and Thomas Quick (from left to right) went to Houston, Texas, to convert a Standard J-1 airplane, which had been built for a four-cylinder, 100-horsepower Hall Scott engine, to receive the 150-horsepower Hispano-Suiza engine that they had bought along with the J-1 from the army under sealed bids. While there, they flew for exhibitions at Ellington Field, as shown in this 1923 photograph. (Author's collection.)

The Quick boys are shown flying the skies above Huntsville during the 1920s. Joe and Thomas visited county fairs in Alabama, making exhibition flights, distributing circulars, and carrying passengers. Curtiss and Erie flew to towns across Alabama looking for passengers. While approaching a landing at Corinth, Mississippi, Curtiss's plane suddenly dove into the ground. The passenger was killed, Curtiss was seriously injured, and the airplane was totally wrecked. (Rick Clark.)

Shown with Curtiss Quick's biplane, which he outfitted with a Super Rhone engine, are, from left to right, Curtiss, Cady Quick, and unidentified. Curtiss designed and constructed the necessary equipment, which he installed in a plane in 1924, for dusting cotton to kill boll weevils. It was probably the nation's first aerial crop duster. He was a barnstorming exhibitionist, and his sister Cady, seated in the biplane, also worked the barnstorming circuit. Curtiss flew emergency medical supplies, cross-country races, and gold bullion from Mexican mines to the mint. While in Houston he was called upon by Clyde Cessna to pilot a plane with him as a passenger and entered the cross-country event from New York to San Diego; they won the race. Later moving to Phoenix, Arizona, he became quite involved as the originator of the use of the "lady bug" as the environmentally correct solution to kill boll weevils and other insects. This was met with a great deal of publicity, including *Readers Digest* and *Colliers* magazines. (Rick Clark.)

Cady Quick, Quick's youngest daughter, took an interest in flying as a very small child, playing on her father's stored airplane. Married at 20 years old to Howard Burns—a pilot at that time and working the barnstormer circuit—Cady, without telling her parents, became an adept "wing walker." She became a pilot at the early age of 23, as one of Alabama's first and youngest female pilots. (Valera McGuire.)

The Quick brothers experimented with various engines they purchased or salvaged: Curtiss OX-5, Hispana-Suiza, Renault, Curtiss OXX-6, Anzani, LeRhone, and different planes, including Canuck (JN4D), Standard J-1, and TM Scout. Erie converted a French LeRhone rotary engine to a static rotary engine with success and formed the Super Rhone Engine and Flying Corporation in Houston. In the photograph, Curtiss (top) inspects an airplane salvage with an unidentified man for a prospective engine and parts. (Rick Clark.)

William Quick's 1908 plane is shown as discovered in 1964 by local chapter 190, Experimental Aircraft Association (EAA), in a shed on the Quick home place, and after moving it out of its 56-year storage. It was in remarkable condition with 85 percent of the plane remaining. Although the engine, gas tank, and radiator were gone, the question remained as to how so much of the wood survived. (Jules Bernard.)

The restoration of the Quick plane was a project of the EAA. The first meeting of the restoration committee was held June 24, 1964, with Rob Maulsby, Chester Klier, and Ed Lamb attending. The EAA undertook the project with much attention to adhering to vintage materials and restoration detail. Some reconstruction activities took place at the old Huntsville Airport on South Parkway as shown in this photograph. (Rob Maulsby.)

After a meticulous, complete restoration, the Quick monoplane was rolled out of the hangar used for reconstruction onto the apron of the old Huntsville Airport. It was positioned and photographed next to a Boeing 727 commercial jet, completing a 50-year connection between the Boeing and Quick families. Shown in the photograph are the Boeing 727 pilot (front center), EAA members, and Lorraine Wicks, Quick's granddaughter (right). (Author's collection.)

Restoration was performed by the EAA with the help of Joe Quick, son of its inventor. Having been initially displayed at the Birmingham Airport, the airplane was returned to Huntsville in 1970 to be exhibited at the Space and Rocket Center. Joe (left front), District 1 Madison county commissioner for 24 years, is shown presenting an original propeller to space center officials, with EAA members present. (Rob Maulsby.)

EAA representatives, Quick family members, and city, county, and Space and Rocket Center officials met at the center on November 4, 1970, to view the monoplane displayed in its place of honor at the world's largest space exhibit. The plane was placed in the missile and space hall, hanging above rocket engines used to power the historical Redstone-Mercury, Jupiter, and Saturn rockets, and next to an X-15 rocket plane famous for flights to the edge of space. The monoplane is currently on display for public viewing in the Davidson Center, shown here, and remains in the county of Madison in which it was built and flown—as preferred by the Quick family. It is no doubt an interesting comparison to the Saturn V moon rocket developed some 60 years later and the space shuttle rocket plane developed some 70 years later, both in Madison County at MSFC. (USSRC.)

Two

HUNTSVILLE-MADISON COUNTY AIRPORT

The World Famous Curtis Biplane People of New York City at Huntsville, Alabama

This advertisement in the Mercury Banner invites the people of Huntsville to an aeroplane exhibition April 6, 1912, at the fairgrounds. (There was no airport). A later article described the event: "Sweeping in wide circles, dipping and soaring and traveling through heavy wind at the rate of a mile a minute, Charles Walsh gave the people of this section their first view of a modern aeroplane in action." (Madison County Records Center.)

Local aviators led by Thomas Quick, and the Jaycees, asked the city to clean off a local field in February 1931, and following completion, it was converted into an airfield. This field, with its sod runways, was known as Mayfair Aviation Field. It is seen in this photograph taken at the dedication, showing a biplane piloted by L. G. Mason of Montgomery—winner of the 125-horsepower race. (Huntsville Public Library.)

Huntsville's new airfield off Whitesburg Drive was formally dedicated Saturday, June 27, 1931, before several hundred Huntsvillians and visitors, as shown in the local newspaper. The dedication program included thrilling races and stunts in which 15 to 20 planes from Alabama, Tennessee, and Georgia took part in a competition for awards donated by Huntsville merchants and the Huntsville Junior Chamber of Commerce. (Madison County Records Center.)

The airfield was "declared by experts to be ideal," according to a June 28, 1931, article in the *Huntsville Daily Times*. Shown are some flyers at the dedication, from left to right, (first row) Jimmy Anderson, Birmingham; Frank Wood, Huntsville; and Hank Prichard, Atlanta; (second row) E. A. Raymond, Birmingham; Bob Baughman, Birmingham; Roy Butler, Montgomery; L. G. Mason, Montgomery; Jim Williams, Atlanta; and George Shealy, Atlanta. The airfield was a modest beginning, and demands soon exceeded capability. (Martel Brett via Kathy Sparks.)

Planes present at the dedication included Waco-F, Byrd, Lincoln, Travelaire, Curtiss Robin, National Guard Falcon, and Stinson. After the dedication address by Mayor McAllister, Lorraine Quick (the niece of Cady Quick, Huntsville's only woman pilot) and the reigning Miss Junior Chamber of Commerce circled the airfield in a Stinson and dropped a quart of wine directly onto the field. (Martel Brett via Kathy Sparks.)

Beneath the summer sun and in a temperature that hovered around 100 degrees Fahrenheit, airfield dedication attendance was sparse. However, the aviation enthusiasts who did brave the fierce rays of the sun were able to fully realize what aviation has come to mean to the world, and how Huntsville would benefit. The August 1931 issue of *Southern Aviation* carried a lengthy story about Huntsville's new 150-acre airfield, located west of Alabama Street between Bob Wallace and Thornton Streets. Shown in this photograph, taken at the airfield dedication, is the contest between a Ford and a Chevrolet. This was part of the daredevil program of aerial stunts, races, and parachute jumping and cavorting of 15 planes in order to draw a crowd. Hal Parker was the master of ceremony for the contest and described the action. The drivers jumped out of the two automobiles just before the head-on crash. (Huntsville Public Library.)

Some of the participants in the Huntsville airfield dedication, from left to right, included Blanche Toney, one of the visiting women fliers, from Birmingham; Lorraine Quick, who christened the new field; and Bob Baughman, manager of the Birmingham Air Service and winner of several events during the dedication. Martel Brett, aviation writer and cameraman, covered the event for the *Birmingham News*. (Martel Brett via Kathy Sparks.)

The original Huntsville airfield was labeled as "landing field" on this 1937 U.S. Geographical Survey (USGS) topographical map and was located south of downtown Huntsville at 34°42' N, 86°34' W. The 1934 Department of Commerce Airport directory described it as a commercial field having four dirt runways with the longest being the 2,400-foot northeast/southwest strip. It closed in 1941. (University of Alabama Map Library.)

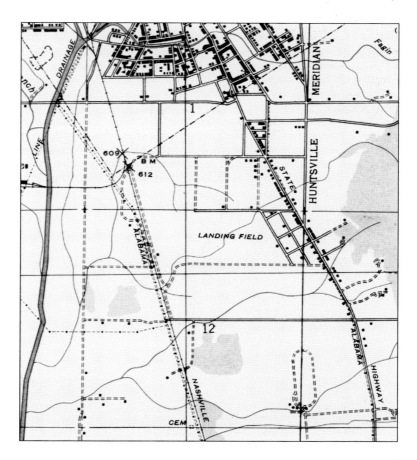

On the night of March 15, 1938, the Huntsville airfield was transformed into a scene of suspense. Eighteen army planes—P-35s of the 17th Pursuit Squadron from Selfridge Field in Michigan on their way to Tampa, Florida, for aerial maneuvers—dropped out of a storm, landing at Huntsville's airfield by automobile lights. Hundreds of automobiles and passengers dashed to the airfield as soon as the WBHP radio station broadcast appealed for lights to help break the darkness and to aid the fliers in their precarious attempts to land. According to a *Huntsville Times* article, the pilots spent the night in the Russell Erskine Hotel. Before the planes took off the next day, Captain Allison expressed his appreciation to the people gathered to see the flyers off. He praised the quick response of the state patrolmen and the Huntsville Police Department, as well as the citizens of Huntsville. (Hubert Williams via James Hill Jr.)

One of the 18 army planes, PA50, is shown in this close-up view at Huntsville airfield. Army plane No. PA70, shown in the previous photograph, is the only known surviving P-35 of World War II, and it is on display in the U.S. Air Force Museum, Wright-Patterson Air Force Base, Ohio. (Hubert Williams via James Hill Jr.)

ORIGINAL CITY AIR MAIL PREPARES 1938 TAKEOFF
. . . Postmaster Collier Instructs Pilot Crofford

On May 15, 1938, a blue and white two-seated Waco JN-4 biplane piloted by John Crawford, shown right, sputtered down the runway of the Huntsville airfield and took off for a one-hour trip to Birmingham. The cargo, weighing only 17 pounds, was the first airmail flight out of Huntsville and is described in this *Huntsville Times* article. L. G. Collier, left, was Huntsville postmaster in 1938. (Madison County Records Center.)

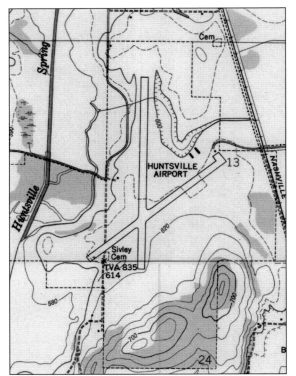

The second Huntsville airport, located in south Huntsville at 34°41' N, 86°35' W, opened in 1941. It had two paved runways with runway 18/36 initially being 4,000 feet long. The terminal building was located at the northeast end of runway 5/23. Airline service started in 1944 by Pennsylvania Central Airline and Eastern in 1946. This 1950 USGS topographical map shows the two paved runways. (University of Alabama Map Library.)

The first Pennsylvania Central Airline (PCA) plane, a DC-3, landed at the Huntsville airport on November 5, 1944. The plane, the *City of Huntsville*, was christened with water from the Big Spring by Susie Spragins, as shown in this photograph. The line was an intrastate commuter service. (Huntsville Public Library.)

J. B. Hill, a local Huntsville retailer and owner of the Jewel Shop in downtown Huntsville, purchased the first ticket for the November 5, 1944, northbound flight of the *City of Huntsville*. Hill is shown boarding the PCA Capitaliner and is also shown in the previous photograph, just behind Susie Spragins christening the plane. (James Hill Jr.)

An Eastern Air Lines flight is shown taxiing out for takeoff amid the waves and cheers of locals with their automobiles lining the runway at this September 1949 air show. The first passenger to buy a ticket on Eastern Air Line's first flight from Huntsville, on January 2, 1946, was Abe Pizitz, a local retailer and Huntsville businessman, who boarded the plane for Chicago. (Huntsville Public Library.)

Shown in this photograph is the first flight from Memphis, Tennessee, to Atlanta, Georgia, of the Eastern Air Lines service in Huntsville. Eastern's headquarters was in New York. Agent Francis Louise Hill, daughter of J. B. Hill, welcomes the crew and passengers onboard the airline's Silver Fleet. (James Hill Jr.)

This September 1950 photograph of the Huntsville-Madison County Airport's original terminal building (around 1944), was taken during a meeting of Eastern Air Lines president Eddie Rickenbacker with 24 of the airline's top corporate officials. The sign at right says, "Air Cargo Terminal, Eastern Air Lines." Little more than a toolshed, the original wooden terminal building was used until construction of a new terminal building in 1951. (Johnny Johnston.)

The role of Huntsville's airport grew during World War II. Postwar, city and county authorities improved the airport, and airlines began adding flights to Huntsville. The airport concession facilities were leased from the city and county by Southern Airways during the 1940s and later by Huntsville Air Service. This photograph shows the original terminal building in 1953 with an Eastern Martin 404 landing on runway 23. (Johnny Johnston.)

The year 1950 marked the start of construction of an administration building; shown here is part of the third phase of development of Huntsville's second airport. Architectural plans for the proposed administration building were drawn up with G. W. Jones and Sons as consulting engineers. With Schrimsher Construction among the contractors, the administration building was completed in 1952. (Huntsville Airport Authority.)

Huntsville's new air terminal was dedicated on April 25, 1953. Senator Sparkman and General Vincent, commander of Redstone Arsenal, dedicated the new air terminal. It can be seen in this 1955 photograph of a Capital DC-3 airliner (PCA changed its name to Capital in 1948). An air traffic control tower was added in the mid-1950s shortly after this picture was taken. (Huntsville Public Library.)

A B-29, the heaviest aircraft to land at the Huntsville airport, set down in July 1955 on a flight to Memphis. Shown is the administration building and several parked planes. By the early 1950s, the community felt that a great deal might come from the airport being operated by an authority, and the airport authority was created in 1956. The airport's air traffic by this time had also improved. (Huntsville Public Library.)

Formerly Pennsylvania Central Airline (PCA), the inaugural flight of Capital Airlines' new Viscount plane service from New York to Huntsville is shown in this September 25, 1956, photograph of the flight crew, Capital Airline representatives, and local Huntsville city and airport officials. (Huntsville Airport Authority.)

This photograph was taken from the airport control tower shortly after it opened on July 25, 1957. The wood structure in the background is the old 1944 airport terminal building. The Whitesburg Drive-in Theater can be seen in the distance. The new Huntsville airport tower was "Traffic Cop of the Airways," according to a July 25, 1957, *Huntsville Times* article. The $370,000 facility operated with eight on staff. (Huntsville Public Library.)

This 1960s photograph was taken looking south from the airport control tower and shows an Eastern Lockheed Electra and a United Viscount awaiting takeoff. Getting on and off the old runway was "like playing football in the bathroom" a veteran commercial pilot cracked between flights. Its approach included mountains, downtown buildings, an open rock quarry, smaller planes, and putting the plane directly over a thickly populated area at low altitude. (Huntsville Public Library.)

This is a photograph taken by Tony Campbell in 1957 of a Convair 440 airplane at the rain-soaked Huntsville airport. Operated by Eastern Air Lines, it was parked at the main gate at that time. (Tony Campbell.)

In this photograph is a Southern Airways DC-3 in July 1961 parked on the west side of the terminal. Incorporated in July 1943, Southern Airways' first DC-3 took off nine years later from Atlanta to Gadsden, Birmingham, and Tuscaloosa. The inaugural run of its jetliners was a decade and a half later in August 1967. (Unidentified photographer via Mike Sparkman.)

This is a photograph of Eastern Air Line's inaugural Lockheed Electra flight in February 1962 in front of the terminal building and control tower at Huntsville-Madison County Airport. During this time, scheduled service was started at that airport with Boeing 727-100s of Eastern and United and DC-9s of Southern. (Huntsville Public Library.)

Shown in this photograph taken on June 13, 1967, is Eastern Air Lines Boeing 727 service at Huntsville-Madison County Airport. Pictured are, from left to right, city and airport officials: Mayor Glenn Hearn, James Record, Richard Hughes, Ed Mitchell, Sonny Stockton, Jack Bentley, and unidentified. (Huntsville Public Library.)

In 1963, the airport authority was reorganized, and plans were announced for the new Jetport, shown here upon completion. The state legislature enacted a bill empowering the authority to issue its own revenue bonds. That bill, according to Ed Mitchell Jr., chairman of the authority during the four-year construction of the Jetport, helped make the Jetport self-supporting without funding from the city and county. (Huntsville Airport Authority.)

The Jetport opened for commercial airline activity on October 29, 1967. The dedication and ribbon cutting was held on September 15, 1968, according to this airport authority event program. It included as special guest speaker the Honorable Alan S. Boyd, secretary of transportation, as well as federal, state, county, and city officials, and a tour of the Skycenter, Hangar, Weather, and FAA and FBO areas. (Huntsville Public Library.)

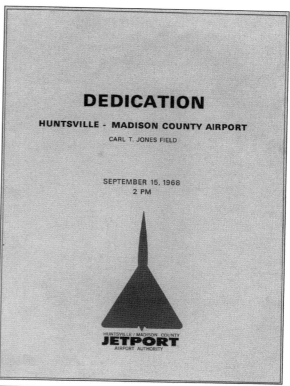

Shown participating in the dedication are, from left to right, (first row) Gen. Charles Eifler, Sen. John Sparkman, Dr. von Braun, and the Honorable Alan S. Boyd, secretary of transportation (speaking). Also participating are Rep. Bob Jones, Gov. Albert Brewer, Madison County commissioners, Madison County chairman James Record, Mayor Glenn Hearn, city councilmen, and the airport authority, consisting of chairman Milton Carter and members Ed Mitchell Jr., Joe Fleming, Henry Bragg, and Jack Bentley. (Courtesy of Sam Tumminello.)

This monument to Carl Jones—"valiant soldier, extraordinary leader, distinguished Alabamian, great American"—is permanently on display at the entrance to the Huntsville Jetplex, after whom the airfield is named in tribute. It was unveiled at the dedication of the Jetport in 1968. (Huntsville Public Library.)

When the Huntsville-Madison County Jetport and Carl T. Jones Field, shown here, opened on October 29, 1967, the old Huntsville airport closed its commercial services simultaneously. A small jet took off from the old airport carrying dignitaries and flew to the new Jetport. This flight marked the last commercial flight out of the old airport and the first flight into the new one. (Huntsville Public Library.)

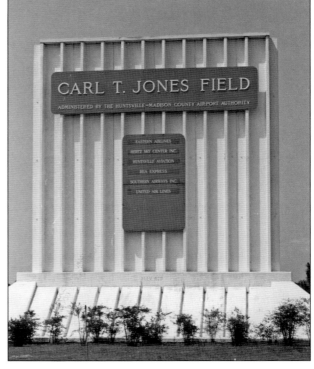

The third Huntsville Airport location was southwest of town at 34°38' N, 86°46' W just off U.S. Highway 20 between Huntsville and Decatur as shown in this 1975 USGS topographical map. The Carl T. Jones 8,000-foot airlines runway, with an extra 2,000 feet available for the giant jetliners of the future: the supersonic transport, jumbo jet, and Airbus. (University of Alabama Map Library.)

When the Huntsville-Madison County Jetport and Skycenter, shown here, were dedicated, it was proclaimed the airport of the 21st century at the ceremony, and some predicted that this pioneer effort would be a model for future airport centers. The first carriers were Eastern Air Lines, United Airlines, and Southern Airlines. (Huntsville Public Library.)

While the third location of the Huntsville Airport southwest of town was opened in October 1967, general aviation, represented here by a Piper plane, continued to use the old Huntsville airport until early 1968 when the Fixed Base of Operations (FBO) facility was finished at the new airport. It was renamed the Huntsville International Airport in 1988. (Huntsville Public Library.)

The International Intermodal Center at the Huntsville International Airport, shown here, transfers business cargo between planes, trains, and trucks. It opened in 1984 and, with the Huntsville International Airport and the Jetplex Industrial Park, comprises the Port of Huntsville. The heart of this unique concept is a 550-ton gantry crane—one cantilever section extending over the railroad staging yard and the other over the truck staging area. (Huntsville Airport Authority.)

Three

REDSTONE ARSENAL

The decade of the 1940s was extremely important to the history of Redstone Arsenal. It was not only the time when the arsenal complex was physically founded but also marked the beginning of the arsenal's modern mission. During the war years, army bombers dropped more than 8 million pounds of munitions at the arsenal. Pictured is an aerial run at a bombing mat in 1945. (AMCOM.)

In July 1941, World War II was sweeping across Europe and Asia. America was re-arming and seeking sites for arms plants. Then Rep. John Sparkman was influential in getting a new chemical munitions plant located southwest of Huntsville. The war department's formal announcement of Huntsville's selection on July 3 is shown here as its site for the chemical munitions manufacturing and storage facility is the subject of the *Huntsville Times* article shown here. (AMCOM.)

The site selected became known as Huntsville Arsenal. In August 1941, the first commanding officer of the Huntsville Arsenal, Col. Rollo Ditto, shown here, arrived and broke ground for the initial construction of the arsenal. Huntsville Arsenal's first production facility was activated in March 1942, just seven months after Colonel Ditto's arrival. The arsenal became the sole manufacturer of colored smoke munitions and also produced gel-type incendiaries and toxic agents. (AMCOM.)

Because of escalating global tensions, Congress had approved funds in April 1941 for the army to construct another chemical manufacturing and storage facility to supplement the production of the Chemical Warfare Service's only chemical manufacturing plant at Edgewood Arsenal. With this funding approval, actual construction on the first permanent building at Huntsville Arsenal, shown in the photograph, began in September 1941. (AMCOM.)

Recognizing the economy of locating an ordnance-loading plant close to Huntsville Arsenal, the chief of ordnance decided to build a facility adjacent to the Chemical Warfare Service's installation for manufacture of artillery shells and other explosives. Initially known as Redstone Ordnance Plant, Col. Carroll Hudson is shown breaking ground for construction in ceremonies held in October 1941. The plant was re-designated Redstone Arsenal in February 1943. (AMCOM.)

Huntsville Arsenal manufactured poison gases such as phosgene and mustard. More conventional output included chemical artillery shells and bombs. Redstone Ordnance Plant produced conventional artillery shells up to 155-mm caliber and rifle grenades. Mass production was perfected for Tetrytol, a high explosive used in bursters and demolition blocks. The first production line, like the one shown, began operation at Redstone Ordnance Plant in March 1942. (AMCOM.)

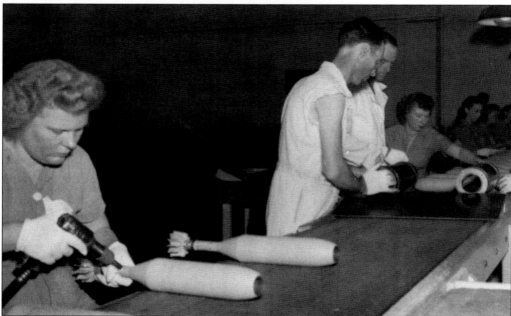

Contrary to existing thought at that time, Col. Hudson began to lay plans for the employment of women in the event that manpower became scarce. The Civil Service Commission shortly announced job examinations for female trainees. At the time, the use of women was frowned upon, but events more than justified Hudson's actions. By December 1942, forty percent of Redstone production line employees, like those shown here, were women. (AMCOM.)

The South Bombing Range was completed in May 1943. By July, a 500-foot concrete bombing mat was finished, and the first skate tests and 4,000-foot drop tests of M50 incendiary bombs were conducted. Also during this period, a simulated village (shown here) was constructed. Known as "Little Tokyo," it was used to test M47 bombs. Later a 200-foot wooden structural target was erected for the proofing of 500-pound M76 incendiary bombs. (AMCOM.)

At the end of the war in 1945, production at both Redstone and Huntsville Arsenals ceased. The ordnance plant was put on standby status in February 1947 while the chemical corps installation was declared excess in September 1947. While Huntsville Arsenal was advertised for sale in July 1949, the army, with the help of Senator Sparkman, found good use for this land—developing a new rocket and missile mission. (AMCOM.)

In October 1949, the secretary of the army approved the transfer of the Ordnance Research and Development Division Sub-Office (Rocket) at Fort Bliss, Texas, to Redstone Arsenal. Among those transferred were von Braun and his team of German scientists who had come to the United States under Operation Paperclip during 1945 and 1946. With the arrival of the Fort Bliss group in 1950, Redstone Arsenal entered the missile era. (AMCOM.)

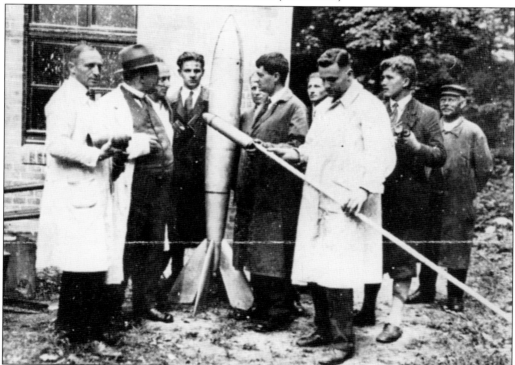

The foremost authority on rocketry outside the United States was Dr. Hermann Oberth (center), a Hungarian-born German. In spring 1930, a young von Braun, second from right, assisted Oberth in his early experiments in testing a liquid-fueled rocket with about 15 pounds of thrust. Years later at the Peenemunde Research Facility in Germany, von Braun's team developed the V-2 rocket, one of the best known of all early missiles. (MSFC.)

This drawing of the German A-4 rocket, forerunner of the V-2, shows its power plant, controls, and propellant tankage. Weighing 27,500 pounds, it was 5.4 feet in diameter, 43.6 feet in length, and had a range of 125 miles. The drawing was copied from documentation on Peenemunde, which was brought to the United States and stored for 12 years at Redstone Arsenal before official return to West Germany in 1958. (MSFC.)

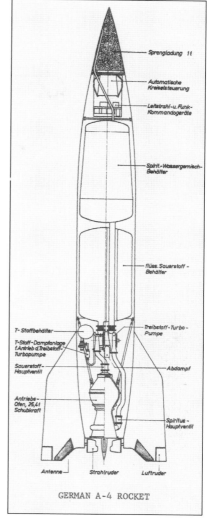

GERMAN A-4 ROCKET

GERMAN A-4 ROCKET COMBUSTION CHAMBER (PRODUCTION)

This photograph of a production version of the A-4 rocket combustion chamber was also contained in the Peenemunde documentation. The A-4 incorporated for the first time a turbo-pump, powered by an 80 percent hydrogen-peroxide steam generator. The engine produced 56,000 pounds of thrust for a burn time of 65 seconds, using Liquid Oxygen (LOX)–Alcohol (75 percent) for its propellant. (MSFC.)

As the war ended, the United States developed an interest in the technical capability of the Germans. A team of American scientists was dispatched to Europe to collect information and equipment related to German rocket progress. Project Paperclip enabled the German rocket specialists to come to the United States to initiate advances in American rocketry. Using a new guidance system (LEV-3) with a 3-axis stabilized platform, some historians estimated that by the end of World War II, the Germans had fired nearly 3,000 V-2 weapons against England and other targets. The German team of specialists was initially assigned to Fort Bliss, Texas, where they reassembled and tested V-2 rockets brought to America from Germany; later they came to Redstone Arsenal. Pictured is a V-2 missile launch at the White Sands Missile Range in New Mexico. (AMCOM.)

In 1947, the Ordnance Corps established—as part of the Hermes Project development of the Bumper research vehicle—a WAC B mated to the German V-2, shown here. In May 1948, Bumper Round 1 was successfully fired at White Sands, the first large two-stage rocket to be launched in the Western hemisphere. Bumper 8, a V-2 and WAC Corporal combination, was the first army missile launched at Cape Canaveral. (AMCOM.)

Project Hermes had officially been started by the U.S. Army in 1944, when General Electric was awarded a contract for design of a family of test vehicles and operational guided missiles. The Hermes program also included the launching of captured German V-2 missiles and the Bumper project. The Hermes, shown here, was first flown successfully in February 1951. It is a liquid-fueled, rocket-powered, surface-to-air missile based on German technology. (MSFC.)

The first definitive contract for U.S. missile research test vehicles was placed with the Jet Propulsion Laboratory/California Institute of Technology in 1944 and led to the tactical Corporal by way of its precursor systems: the Private, development Corporals, and Bumper WAC. Technical supervision of the overall program rested with the Jet Propulsion Laboratory until 1955, when Redstone assumed this mission. The embryo of the army missile programs, the Corporal was a surface-to-surface guided missile, which could deliver a nuclear or high-explosive warhead up to a range of 75 nautical miles. The first Corporal battalion was deployed in Europe in 1955. This system remained in the field until 1962–1963, when it was replaced by the Sergeant missile system. An operational Corporal missile system is shown in this photograph on display at Armed Forces Day on the square in downtown Huntsville in 1957. (AMCOM.)

In December 1955, the U.S. Department of the Army formally announced the establishment of the Army Ballistic Missile Agency (ABMA). The agency would be under the command of Maj. Gen. John Medaris. In January 1956, a ceremony was held in front of Redstone Arsenal headquarters, shown here, honoring his official arrival to assume command. (AMCOM.)

Colonel Hudson, commanding officer at Redstone Arsenal, officially established the Provisional Redstone Ordnance School. In March 1952, he became the first commandant of the school, the forerunner of the Ordnance Guided Missile School. Early training at the school was on the "Loon," shown here, an American-built version of the V-1 Buzz Bomb used to bomb London. (AMCOM.)

When scientists at the Jet Propulsion Laboratory discovered that Thiokol's polymers made ideal rocket fuels, Thiokol moved into the new field, opening laboratories at Elkton, Maryland. In 1949, the Elkton Division signed an army contract to research and develop rocket propellants. This activity moved to Redstone Arsenal, where it began operation in June 1949. Located on-site, the Thiokol area is shown in this 1959 aerial photograph. (AMCOM.)

The Honest John, shown here, was a spin-stabilized, free flight rocket capable of delivering a nuclear warhead. This highly mobile system was designed to fire like conventional artillery in battlefield areas. The Office of the Chief of Ordnance (OCO) assigned Redstone Arsenal responsibility for the preliminary design study for the special purpose, large caliber field artillery rocket in 1950. The basic Honest John system was first deployed in 1954. (AMCOM.)

During the final months of World War II, U.S. defense contractors studied the use of guided missiles to intercept bombers and surface-to-surface missiles. One of these projects, Nike, shown here, would grow to a full deployment of more than 240 missile sites in the United States. The OCO assigned Redstone Arsenal responsibility for supervising and coordinating the research and development phase of the Nike program in August 1951. (AMCOM.)

Although war missiles were the official business, von Braun and the other German scientists who came to Huntsville in 1950 envisioned using the same technology to send men to the moon and other planets. For an early 1950s interview, von Braun poses with a model of a manned Mars ship he designed. (MSFC.)

The OCO formally transferred research and development responsibility for the Redstone project to Redstone Arsenal in July 1951. Known by various names such as Hermes Cl and Major, the OCO officially assigned the Redstone missile its name in April 1952. Also known as the army's Old Reliable, shown here, the Redstone was capable of transporting nuclear or conventional warheads against targets at ranges up to approximately 200 miles. (MSFC.)

A Redstone missile is shown being test fired at the Redstone test stand in the early 1950s. A high-accuracy, liquid-propelled, surface-to-surface missile developed by the von Braun team under U.S. Army management, the Redstone was the first major rocket development program in the United States. The U.S. Department of the Interior's National Park Services designated the Redstone Test Stand as a National Historic Landmark in 1986. (MSFC.)

Pictured is the first Redstone missile flight test at Cape Canaveral, Florida, in August 1953. The first Redstone fabricated and assembled by the Chrysler Corporation was flight tested at the Cape in July 1956. The first successful troop launching of a tactical Redstone occurred at the Cape in May 1958. A Redstone was fired to an altitude in excess of 200,000 feet and a nuclear device of a megaton was detonated in July 1958—the first such accomplishment by the United States. The operational Redstone's range was 57.5 to 201 miles, utilizing a Rocketdyne North American Aviation (NAA) 75-110 A-7 78,000-pound thrust engine burning ethyl alcohol and liquid oxygen. It was controlled by a Ford Instrument Company ST-80 inertial guidance system and was steered by carbon jet vanes, air rudders, air vanes, and spatial air jet nozzles, reaching Mach 5.5 at reentry. (MSFC.)

The first true-engineered thesis for a minimum satellite vehicle using existing Army Ordnance Corps hardware was published in September 1954. Written by von Braun and his team shown here, it proposed using the Redstone missile as the main booster of a four-stage rocket for launching artificial satellites. The plan was later expanded into a joint army-navy proposal called Project Orbiter. However the nation pursued the navy's Project Vanguard instead (AMCOM.)

The Hawk, a medium range, surface-to-air guided missile, provided air defense coverage against low-to-medium-altitude aircraft. The missile was highly lethal, reliable, and effective against electronic countermeasures. Basic Hawk was developed in the 1950s and initially fielded in 1960. In June 1956, the first guided Hawk missile hit the nose of an F-80 drone flying at 11,000 feet and demolished it (shown here) in a shower of flaming fragments. (AMCOM.)

Maj. Gen. John Medaris (left), commander of the Army Ballistic Missile Agency, and Maj. Gen. Holger Toftoy (right), commander of Redstone Arsenal, dedicated this historical marker located at Memorial Parkway and Airport Road. It reads: "Hermes Guided Missile —First American-made guided missile put on public display. First showing was May 14, 1953 at Huntsville, home of the world's largest rocket and guided missile research and development center, Redstone Arsenal." (AMCOM.)

Construction of the largest static test stand in the United States for testing rocket motors was completed at Redstone Arsenal in August 1956. It was slated for use in the Jupiter program, an Intermediate Range (1,500 mile) Ballistic Missile (IRBM). The Jupiter buildup effort included ancillary test area buildings, guidance and control laboratory, and an engineering building called 488, later renumbered 4488, which became the ABMA headquarters building. (AMCOM.)

Jupiter C Missile RS-27 is shown in assembly at ABMA. The Jupiter C was a modification of the Redstone and originally developed as a nose cone reentry test vehicle for the Jupiter IRBM. In September 1956, RS-27 attained an altitude of 682 miles and achieved a velocity of Mach 18, enough to have put its fourth stage into orbit if permission had been granted to do so. (MSFC.)

Jupiter Missile AM-1, shown here, was successfully fired by ABMA in May 1957. This demonstration marked the initial U.S. flight of an IRBM. Despite this achievement, the army's responsibility remained limited to missiles having ranges of 200 miles or less. The Soviet Union's successful launch of *Sputnik 1* in October 1957 resolved this problem by prompting President Eisenhower to approve development of both the army's Jupiter and the navy's Thor. (AMCOM.)

In this October 1957 photograph, von Braun is shown displaying the first nose cone to be recovered from outer space. It was launched on Jupiter C Missile RS-40 in August 1957. In November, the secretary of defense ordered ABMA to prepare a Jupiter C missile to launch a satellite as part of the International Geophysical Year (IGY) program, after the Soviet Union's successful launch of *Sputnik 1* in October 1957. (AMCOM.)

This photograph was taken of von Braun and his associates in 1958 at ABMA, in front of building 4488 on Redstone Arsenal. Pictured from left to right are Dr. Ernst Stuhlinger, Helmut Hoelzer, Karl Heimburg, Ernst Geissler, Erich Neubert, Walter Haeussermann, von Braun, William Mrazek, Hans Hueter, Eberhard Rees, Kurt Debus, and Hans Maus. (MSFC.)

The Army Ordnance Missile Command (AOMC) was created in March 1958. The command's subordinate elements included ABMA, Redstone Arsenal, Jet Propulsion Laboratory, and White Sands Proving Ground. Major General Medaris became its first commanding general. Major General Toftoy became its deputy commander. In November 1958, AOMC headquarters ended its joint occupancy of the ABMA headquarters building 4488, shown here, when it opened for business in its new building 4505. (AMCOM.)

The AOMC also carried out the army's outer space projects for the Advanced Research Projects Agency (ARPA). After AOMC and ARPA signed a memorandum of understanding in September 1958, the Saturn program began at ABMA under army management. Saturn design studies were authorized to proceed at Redstone Arsenal for development of a 1.5-million-pound thrust, clustered engine first stage, shown here. (AMCOM.)

64

Jupiter C Missile RS-29 is shown ready for launch at Cape Canaveral, Florida, on January 31, 1958. It placed *Explorer 1*, the first U.S. satellite, into orbit around the earth. ABMA, in cooperation with the Jet Propulsion Laboratory (JPL), developed and launched the rocket. The Jupiter C vehicle consisted of a modified Redstone missile with two solid-propellant upper stages. The tanks of the Redstone were lengthened and the instrument compartment was smaller. The second and third stages were clustered in a tub atop the vehicle consisting of an outer ring of 11 scaled-down Sergeant rocket engines and a cluster of three scaled-down Sergeant rockets grouped within. The Juno I satellite launch vehicle added a fourth stage atop the Jupiter C's third stage and was fired after third stage burnout to boost the satellite to an orbital velocity of 18,000 miles per hour. (AMCOM.)

After the Russian *Sputnik 1* was launched in October 1957, the launching of an American satellite assumed much greater importance. After the navy's Vanguard rocket exploded on the pad in December 1957, a modified Redstone rocket lifted the first American satellite into orbit just three months after the von Braun team received the go-ahead. The *Huntsville Times* newspaper carried the headline "Jupiter-C Puts Up Moon." (AMCOM.)

Shown here from left to right, chamber president Jimmy Walker, Mayor R. B. Searcy, Chief of AOMC Stuart Jones (behind), and Huntsville Industrial Expansion Committee chairman Dorsey Uptain are celebrating the launch of *Explorer 1* in downtown Huntsville on January 31, 1958, with their own fireworks show. National headlines featured the three key men responsible for the success of *Explorer 1*—Dr. William H. Pickering, Dr. James A. van Allen, and Dr. Wernher von Braun. (AMCOM.)

In this photograph, the *Pioneer IV* is shown being launched in March 1959 on a Juno II rocket at the Atlantic Missile Range, Cape Canaveral Air Force Station. As a joint ABMA/JPL project under the direction of NASA, it achieved a velocity in excess of 24,560 miles per hour, passed within about 36,000 miles of the moon, and traveled on to become the first U.S. satellite in permanent orbit around the sun. Radio contact with *Pioneer IV* was maintained to a record distance of 406,620 miles. The Juno II was derived from the Jupiter missile, which was used as the first stage. Sergeant rocket motors were used as upper stages—11 for the second stage, 3 for the third stage, and 1 for the fourth stage—the same configuration as used for the upper stages of the smaller Juno I rocket. (AMCOM.)

The flight of monkeys Able and Baker, shown here, in May 1959 marked the first successful recovery of living creatures after their return to earth from outer space. The monkeys rode in the nose cone of Jupiter Missile AM-18 to an altitude of 300 miles and a distance of 1,500 miles. Their survival at speeds over 10,000 miles per hour was the first step toward putting a man into space. (AMCOM.)

U.S. Army Missile Command (MICOM) was activated in 1962 and AOMC was discontinued. The Pershing I, shown here, was deployed the following year, replacing the Redstone. The Pershing II's increased range and pinpoint accuracy were major factors influencing the Soviet Union's decision to seek the Treaty on Intermediate Range Nuclear Forces, in which the United States and the Soviet Union agreed to eliminate an entire class of nuclear missiles. (AMCOM.)

In 1962, the Deputy Commanding General (DCG), Land Combat Systems, was responsible for Missile B, Pershing, Sergeant, Redstone, Corporal, Honest John, Little John, Lacrosse, SS-11, 2.75-Inch Rocket, Light Antitank Weapon (LAW), TOW, ENTAC, SS-10, and Shillelagh. The DCG, Air Defense Systems, was responsible for Nike Zeus, Field Army Ballistic Missile Defense System (FABMDS), Hawk, Nike Hercules, Mauler, Nike Ajax, Redeye, and Target Missiles/Multisystem Test Equipment (MTE). Shown here are seven of these weapon systems. (AMCOM.)

In June 1960, a Nike Hercules guided missile (left) tracked and shot down a Corporal ballistic missile (right)—the first time a missile killed a missile. In 1962, a Nike Zeus and a Titan II etched the famous X in the South Pacific sky proving a missile could intercept an ICBM. The June 1984 landmark impact kill of a Minuteman I reentry vehicle by the Homing Overlay Experiment demonstrated a bullet could hit a bullet. (AMCOM.)

The Surface-to-Air Missile, Development (SAM-D) Missile System was renamed the Patriot Air Defense Missile System in May 1976. The name change marked the system's entry into full-scale development. Capable of defeating both high performance aircraft and tactical ballistic missiles, Patriot, shown here, is the only operational air defense system that can shoot down attacking missiles. The first combat use of Patriot occurred in Saudi Arabia and Israel during Operation Desert Storm. (AMCOM.)

First used in combat in Panama in 1989 and later in the Middle East, the Apache helicopter is shown with its 30-mm gun and a mixture of Hellfire missiles and Hydra 70 rockets. Relocation of the aviation troop command from St. Louis, Missouri, to Redstone Arsenal and subsequent establishment of the Aviation and Missile Command (AMCOM) in October 1997, consolidated the management of army aviation platforms and their armament missiles at Redstone Arsenal. (AMCOM.)

Four

REDSTONE AIRFIELD

In 1943, the 6th Army Air Forces (AAF) Base Unit, Proving Ground Detachment, was stationed at Huntsville Arsenal. Consisting of three officers, three enlisted men, and two planes—a B-26 and an L-20—the unit's mission was to support incendiary proofing being conducted by the arsenal's inspection division. An airstrip, shown here, was built on the arsenal to accommodate the planes used to test clusters of incendiary bombs and smoke grenades. (AMCOM.)

Ceremonies marking the opening of the renovated and expanded Redstone Army Airfield were held in July 1956. A light plane shown here "riding a beam"—the first sent out from the recently erected air transportation control tower at the ABMA airstrip—skimmed down and cut a gaily fluttering ribbon strung across the runway to officially activate the new area control system. (AMCOM.)

Shown participating in the airfield dedication ceremony are, from left to right, Major General Medaris, ABMA commander; R. B. Searcy, mayor of Huntsville; and Brigadier General Toftoy, Redstone Arsenal commander. Carrying the ABMA commander, the small aircraft—the first plane to land at the newly renovated airfield—cut the ceremonial ribbon with its propeller. Shortly thereafter, the ABMA commander completed the dedication ceremony by snipping a ribbon on the newly opened control tower. (AMCOM.)

The Redstone Army Airfield, shown on this 1950 USGS topographical map, is located west of Huntsville on Redstone Arsenal at 34°40' N, 86°41' W. Recognizing the need for airlift—since commercial airline service out of Huntsville was severely limited at the time—and for transportation of classified material on military aircraft, General Medaris inaugurated the army service with two pilots and one L-23 aircraft. (University of Alabama Map Library.)

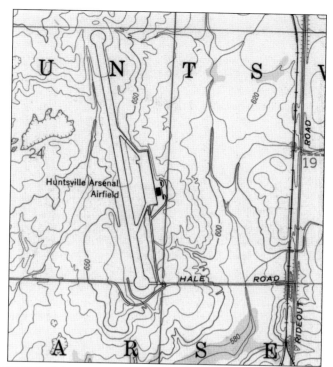

The command's airlift capabilities were improved in 1956 by the addition of an L-20 cargo plane and a DHC-3 Otter. By 1959, Redstone Army Airfield was assigned a fleet of seven twin-engine planes, one single-engine cargo plane, and one helicopter. The airfield's runway and hangar layout is shown in this 1962 photograph. It is the only army airfield with a NASA tenant—the MSFC. (Huntsville Airport Authority.)

The chief of ordnance approved funds in 1961 to be used to construct a closed hangar. Overall, Redstone's aviation activity was outstanding, but maintenance facilities were considered deficient because of the large, open-ended hangar, shown in this 1960s photograph. By 1968, there were four hangars lining the airfield's taxi strip, two used by airfield personnel, one belonging to MSFC, and one supporting the missile command's research and development efforts. (AMCOM.)

Elected to the U.S. Senate in 1948, John Sparkman (center) played an influential role in convincing the army to consolidate its new missile and rocket research efforts at Redstone Arsenal. In later years, he visited Redstone often, always supportive of the army's efforts in the field of missilery. He is seen here with Major General Toftoy (far left) and Major General Medaris (far right) in December 1956, touring missile facilities. (AMCOM.)

A total of four sitting U.S. presidents have landed in *Air Force One* at Redstone Army Airfield. In addition, the facility has supported the visit of another chief executive who landed at the Huntsville-Madison County Jetplex. The first to visit was President Eisenhower (center), who came to Redstone in September 1960 to dedicate NASA's MSFC. (AMCOM.)

Members of the U.S. House of Representatives Subcommittee on Manned Space Flight visited MSFC in March 1962 to gather firsthand information on the nation's space exploration program. The congressional group was briefed on MSFC's manned space efforts earlier in the day and then inspected mockups of two post-Apollo projects. Von Braun bids farewell to Texas Democratic representative Olin Teague, left, before departure at the Redstone Arsenal airstrip. (MSFC.)

President Kennedy, center, came to the arsenal for the first time in September 1962 for a two-day visit to compile firsthand information on the progress of America's space program, shown here with Vice President Johnson, right, and Redstone Arsenal and MSFC officials. Kennedy returned to Redstone in May 1963, where he delivered a speech at the airfield, the first presidential speech made at the arsenal. (MSFC.)

Two U.S. congressmen, accompanied by the NASA administrator, visited MSFC in April 1964 for a briefing on the Saturn program and a tour of the facilities. Shown are, from left to right, Congressman Gerald Ford Jr., Republican representative of Michigan; von Braun, MSFC director; Congressman George Mahon, Democratic representative of Texas; and James Webb, NASA administrator. (MSFC.)

George Wallace Jr., center, was governor of Alabama for four terms, spanning the years 1963 to 1987. He ran for president four times and is shown here at Redstone Army Airfield in June 1971 at the beginning of his second gubernatorial term, and a year before an assassination attempt on his life in 1972 while running for president. (AMCOM.)

Richard Nixon, left, visited Huntsville in 1974 to be the main speaker at the second annual Honor America Day celebration. He arrived at Redstone Army Airfield and then motored to Big Spring Park in downtown Huntsville. Although Ronald Reagan did not land at the arsenal, the installation's airfield was put on 24-hour alert status to support the presidential flight detachment on his visit to Decatur, Alabama, in 1984. (AMCOM.)

In June 1990, Pres. George Bush, left, came to Huntsville to speak at a Republican campaign luncheon downtown and to make some remarks at MSFC. He and Alabama governor Guy Hunt, center, are greeted by MSFC director Jack Lee upon their arrival at Redstone Arsenal Airfield. This was the first sitting president to visit MSFC since President Kennedy's visit almost 30 years earlier. (MSFC.)

Three of the army pilots who have operated out of Redstone Army Airfield in the past were Maj. Gen. John Medaris, Maj. Gen. John Barclay, and Maj. Gen. August Cianciolo; they constitute Redstone's trio of flying generals. In June 1957, Medaris, who received his silver pilot's wings from the Army Aviation School at Fort Rucker, Alabama, was one of the army's highest-ranking officers to complete flight training. (AMCOM.)

Barclay, ABMA commander from 1958 to 1960 as well as deputy commanding general of AOMC from 1960 until his retirement in June 1961, was the second of Redstone Arsenal's highest ranking officers to win his pilot's wings. After he graduated from the Army Aviation School at Fort Rucker in 1958, he became one of the most active pilots on the airfield's roster of aviators. (AMCOM.)

The last of Redstone Arsenal's flying generals, Cianciolo served as MICOM commander from July 1988 to September 1989. Unlike the arsenal's other two high-ranking aviators, Cianciolo earned his pilot's wings before he was assigned to Redstone Arsenal. His broad background as an army aviator included combat experience in Vietnam. Among his decorations are the Air Medal with "V" device (for valor) and the Master Army Aviator Badge. (AMCOM.)

Von Braun was the second of three sons born to Baron von Braun and Baroness Emmy von Quistorp. His father, center, is shown upon arrival to visit his son in September 1959. Accompanying Baron von Braun is Curt Jurgens, left, a successful European film actor. Curt, born in Solln, Bavaria, would abandon the country of his birth after the end of World War II and became an Austrian citizen in 1945. (AMCOM.)

Gen. Charles Eifler's career included three tours at Redstone Arsenal. From July 1959 until August 1961, he was commandant of the U.S. Army Ordnance Guided Missile School. General Eifler then returned to Redstone Arsenal as deputy commanding general, Land Combat Systems, MICOM. He then served as the MICOM commanding general until 1969, shown here, far left, reviewing troops at Redstone Airfield. (AMCOM.)

The Guppy was built by John Conroy of Aero Spaceliners, Inc., who started with the fuselage of a surplus Boeing C-97 Stratocruiser and ballooned out the upper deck, providing a 25-foot clear diameter. The front section hinged so that it could be folded back 110 degrees. NASA used the aircraft to transport the S-IVB upper stage used on Saturn V launch vehicles, between West Coast and southeast facilities. (MSFC.)

This photograph was taken at the Redstone Arsenal Airfield during the unloading of the S-IVB stage from the Guppy. The stage housed the Orbital Workshop (OWS), which measured 22 feet in diameter and 48 feet in length. The S-IVB stage was modified at the McDonnell Douglas facility at Huntington Beach, California. The Guppy was later used to transport flight articles for the International Space Station. (MSFC.)

The space shuttle orbiter *Enterprise*, shown riding piggyback on its Boeing 747 carrier jet, flew over Huntsville and landed at Redstone Arsenal Airfield in 1978. Approximately 85,000 people visited the *Enterprise* on display at Redstone Arsenal. NASA uses two extensively modified Boeing 747 airliners to transport space shuttle orbiters. One is a 747-100 model, while the other is a short range 747-100SR. (MSFC.)

The *Enterprise* is shown being off-loaded at Redstone Arsenal Airfield for later Mated Vertical Ground Vibration tests (MVGVT) at MSFC's Dynamic Test Stand. The orbiters are placed on top of the Shuttle Carrier Aircraft (SCA) by Mate-Demate Devices, large gantry-like structures, which hoist the orbiters off the ground for post-flight servicing and then mate them with the SCAs for ferry flights. (MSFC.)

Five

Marshall Space Flight Center

President Eisenhower and Katherine Marshall, widow of Gen. George Marshall, are shown unveiling the bronze bust of her late husband during the dedication ceremony of MSFC on September 8, 1960. As a new field installation of NASA formed within the boundaries of Redstone Arsenal, MSFC began its operation on July 1, 1960, after a transfer ceremony, with von Braun as center director. (MSFC.)

The AOMC and its ABMA transferred its space-related missions, along with about 4,000 civilian employees and $100 million worth of buildings and equipment at Redstone Arsenal and Cape Canaveral to NASA's MSFC. Von Braun and Maj. Gen. August Schomburg officiated this July 1, 1960, transfer ceremony, shown here taking place in front of the ABMA-MSFC joint headquarters, building 4488. (MSFC.)

This photograph was taken September 30, 1961, at the official groundbreaking ceremony for a new headquarters building at NASA's MSFC. Joining von Braun, left, in the ceremony was Sen. Robert Kerr, then chairman of the Senate Committee on Aeronautics and Space Sciences. (MSFC.)

Astronaut Alan Shepard Jr. is shown lifting off in the *Freedom 7* Mercury spacecraft on May 5, 1961. The Project Mercury mission (MR-3), using a Mercury-Redstone vehicle developed by the von Braun rocket team, was the first manned space mission for the United States. During the 15-minute suborbital flight, Shepard reached an altitude of 115 miles and traveled 302 miles downrange. His flight lasted 15.5 minutes. Shepard became the first American in space as a result of this mission. The Mercury-Redstone vehicle was designed, built, and the mission managed by MSFC. Redstone missiles were modified for NASA's Mercury program. The propellant tanks were elongated by 96 inches, adding 20 seconds of burn time. The warhead section of the missile was replaced with the Mercury capsule and escape tower. It used the same engine—the 78,000-pound, LOX/Ethyl Alcohol Rocketdyne A-7. (MSFC.)

Von Braun is shown addressing a crowd celebrating in front of the Madison County Courthouse following the successful launch of astronaut Alan Shepard—America's first astronaut in space—on a Mercury-Redstone Launch Vehicle. Shepard's Mercury spacecraft was launched from Cape Canaveral. Three weeks later, President Kennedy proposed that U.S. astronauts go to the moon within the decade. (MSFC.)

President Kennedy (right) visited MSFC on September 11, 1962. Here he and von Braun, MSFC director, are shown touring one of the laboratories. President Kennedy came to talk with von Braun and the men who would build the rockets to make true his commitment to put men on the moon "before this decade is out." (MSFC.)

In the foreground of this 1959 photograph of Huntsville rocketry is the Redstone vehicle that, as the Jupiter C vehicle configuration, orbited *Explorer 1* and, as the Mercury Redstone vehicle's booster, launched the first Americans on brief suborbital flights into space. Behind it is the Juno II vehicle, a modified Jupiter used to launch satellites and the first moon probes. Dwarfing both is the Saturn I vehicle, an eight-engine giant. The vehicle and its later Saturn IB version flew perfectly on all 19 flights. More than any of the Saturn vehicles, the Saturn I S-I first stage configuration evolved during flight tests. The first four launches used the Block I vehicle with inert upper stage. Block II versions carried a live upper stage, the S-IV. The S-I first stage for the Saturn I also became the first stage of the Saturn IB. In this application it was called the S-IB. (MSFC.)

The Saturn I S-I stages for the SA-4, SA-6, and SA-7 missions are shown being assembled at the fabrication and assembly engineering division in MSFC building 4705 on January 13, 1963. A central 2.67-meter diameter liquid oxygen tank—from Jupiter—was surrounded by eight 1.78-meter outer tanks—from Redstone—used alternately for liquid oxygen and kerosene. (MSFC.)

Shown in this photograph is a static test firing of the Saturn I S-I stage (first stage) at MSFC during February 1961. The Saturn I is the first of the Saturn launch vehicle family. The Saturn I S-I stage (first stage) had eight H-1 engines clustered, using liquid oxygen/kerosene-1 (LOX/RP-1) propellants capable of producing a total of 1.5 million pounds of thrust. (MSFC.)

MSFC's first Saturn I vehicle, SA-1, is shown lifting off from Cape Canaveral, Florida, on October 27, 1961. This early configuration, Saturn I Block I, which is 162 feet tall and weighs 460 tons, consisted of the eight H-1 engines S-I stage and the dummy second stage (S-IV stage). (MSFC.)

Workers at MSFC are shown hoisting S-IB-1, the first flight version of the Saturn IB launch vehicle's first stage (S-IB stage), into the Saturn IB static test stand on March 15, 1965. Developed by MSFC and built by the Chrysler Corporation at the Michoud Assembly Facility (MAF) in New Orleans, Louisiana, the 90,000-pound booster utilized eight H-1 engines to produce a combined thrust of 1.60 million pounds. (MSFC.)

S-IB-1, the first flight version of the Saturn IB launch vehicle's first stage (S-IB stage), is shown undergoing a full-duration static firing in the Saturn IB static test stand at the Marshall Space Flight Center (MSFC) on April 13, 1965. Between April 1965 and July 1968, MSFC performed 32 static tests on 12 different S-IB stages. The first stage engines of the Saturn I, Saturn IB, and Saturn V (respectively the S-I, S-IB, and S-IC stages) used a non-cryogenic fuel called RP-1, derived from kerosene. All Saturn's engines used liquid oxygen as the oxidizer. As in so many engineering achievements, engine development for the Saturn program represented the culmination of earlier research and development efforts. The S-IB's H-1 engine traces its lineage to no less than five prior designs, with major components derived from hardware applied in the Thor, Jupiter, and Atlas engines. (MSFC.)

SA-201, the first Saturn IB launch vehicle developed by MSFC, is shown lifting off from Cape Canaveral, Florida, on February 26, 1966. This was the first flight of the S-IB and S-IVB stages, including the first flight test of the liquid hydrogen/liquid oxygen–propelled J-2 engine in the S-IVB stage. During the 37-minute flight, the vehicle reached an altitude of 303 miles and traveled 5,264 miles downrange. (MSFC.)

The 1960 transfer of the von Braun team from the army's ABMA to NASA's MSFC involved land, facilities, and personnel. ABMA's guidance and control complex, shown in the foreground of this aerial photograph, became MSFC's Astrionics Laboratory. ABMA's fabrication and assembly facilities, shown in the background, transitioned to MSFC's Manufacturing Engineering, Test, and Data Systems Laboratories. (MSFC.)

ABMA's structures and mechanics facility, shown in this aerial view, transitioned to MSFC's Propulsion and Vehicle Engineering Laboratory with its structures, propulsion, materials, dynamics, and systems functions. ABMA's test facilities, shown on the horizon, expanded into the MSFC Test Laboratory's complex of three large test stands for engine, dynamic, and booster testing. (MSFC.)

A complete F-1 engine assembly is shown in this photograph. Designed and developed by Rocketdyne under the direction of MSFC, the engine measured 19 feet by 12.5 feet at the nozzle exit, and each engine produced a 1.5-million-pound thrust using liquid oxygen and kerosene as the propellant. (MSFC.)

MSFC personnel checked out the first batch of production F-1 engines during 1963, sending the F-1's thundering roar through the Tennessee River valley. The first F-1 rocket engine arrived at MSFC on October 29, 1963. Pictured is the F-1 engine undergoing a Static Test Firing at the S-IB Static Test Stand. The engine was the most powerful in the world, and it would require five of them to power the Saturn V. Although the F-1 had its roots in early air force studies, it was a newer engine than the Saturn I's H-1 engine. Both used liquid oxygen and RP-1 propellants, but size and performance characteristics made the F-1 fundamentally different. Basic research, development, and manufacturing took place at Rocketdyne facilities in Canoga Park, California, and many component tests were conducted at the company's Santa Susana Field Laboratory in the mountains nearby. (MSFC.)

Fuel and LOX tanks are shown being built in Huntsville for the Saturn V first stage, S-IC. In the fabrication of the 10-meter (33 foot) diameter booster tankage, new tooling of unique size and capabilities had to be built, and fabrication of the tank cylinders and domes required circumferential welds and meridian welds of unprecedented length. These had to pass stringent inspection to "man-rate" the Saturn V vehicle. (MSFC.)

The fuel tank assembly of S-1C-T (the first stage of the Saturn V test vehicle) is shown being mated to the LOX tank at MSFC. The two tanks were connected by a 26-foot intertank section. Other parts of the booster included the forward skirt and the thrust structure, on which the five F-1 engines were to be mounted, each weighing 10 tons. (MSFC.)

The S-IC-T stage is shown being hoisted into the S-IC Static Test Stand at MSFC. The S-IC-T stage was a test vehicle that was ground tested repeatedly over a period of many months to prove the vehicle's propulsion system. The 280,000-pound stage, 138 feet long and 33 feet in diameter, housed the fuel and liquid oxygen tanks that held a total of 4.4 million pounds of liquid oxygen and kerosene. (MSFC.)

This photograph shows the intense smoke and fire created by the five F-1 engines from a test firing of the Saturn V first stage (S-1C) in the S-1C test stand at MSFC. The next day, April 16, 1965, MSFC personnel successfully test fired all five of the S-IC-T stage's F-1 engines, lasting 6.5 seconds. This first S-IC-T five-engine test occurred two months ahead of schedule. (MSFC.)

S IC FLIGHT STAGE–HUNTSVILLE

IND B1200-74

This small group of officials is dwarfed by the gigantic size of the Saturn V first stage (S-1C) in this photograph at the shipping area of the Manufacturing Engineering Laboratory at MSFC. S-IC flight stages S-IC-1, S-IC-2, and S-IC-3 were fabricated and assembled at MSFC in the Manufacturing Engineering Laboratory. Each was successfully acceptance tested by MSFC's Test Laboratory, completing a highly successful Saturn V first stage development and acceptance test program. The S-IC stages were towed through MSFC on their way to the adjoining Tennessee River and their barge transportation to Cape Canaveral, Florida, using the Poseidon, an oversized barge built to carry the big S-IC boosters of the Saturn V. The stages were transported down the Tennessee, Ohio, and Mississippi Rivers across the Gulf of Mexico and through Florida's intracoastal waterway to the Kennedy Space Center. (MSFC.)

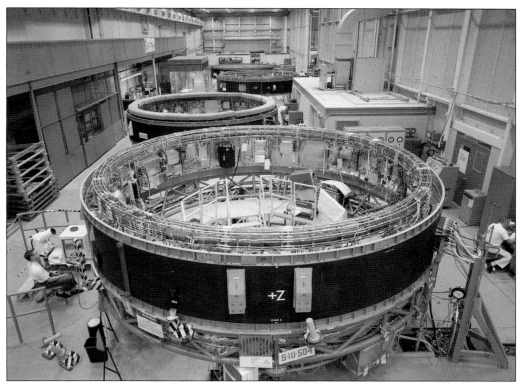

The brain and nerve center of the Saturn V, the Instrument Unit (IU), is shown in production at IBM's Huntsville facility in Research Park. The basic functions of the IU included guidance and control, command and sequence, relay of data, and stabilization. It evolved as an in-house project at MSFC and was based on the guidance expertise accumulated from the V-2, the Redstone, and subsequent vehicles. (MSFC.)

NASA reached a critical point in the Apollo program with the *Apollo 4* flight (SA-501) on November 9, 1967, an all-up launch from launch complex LC-39 at Kennedy Space Center—perhaps the biggest gamble of the Apollo-Saturn program. The flight, shown here, was termed "perfect" based on evaluation of flight data and circumvented the costly and time-consuming process of incremental flight testing of each stage prior to launching a complete vehicle. (MSFC.)

On July 16, 1969, *Apollo 11*, shown here, was launched. Most of the American public and the world knew the towering 111-meter rocket as the Saturn V or the Apollo 11 mission. To the men and women who built it, it was known better by its official designation: SA-506. Whatever its name, everyone knew its destiny. This rocket was going to be the first to land men on the moon. The number of observers around the launch site was conservatively estimated to be a million, including congressmen, ambassadors, governors, mayors, and other public figures. Vice Pres. Spiro Agnew, former Pres. Lyndon Johnson, and Lady Bird Johnson were there. Live television coverage of the liftoff was beamed across six continents, and there were an estimated 25 million viewers in the United States alone. Additional millions heard the event by radio. (MSFC.)

Apollo 11 astronaut Edwin Aldrin Jr. is shown next to the Lunar Module (LM) he piloted on the first manned lunar landing mission. Aboard the spacecraft was astronaut Neil Armstrong, commander. The Command Module (CM), piloted by Michael Collins, remained in a parking orbit around the moon while the LM, named Eagle, carrying astronauts Armstrong and Aldrin, landed on the moon in the Sea of Tranquility on July 20, 1969. (MSFC.)

900956   1969
Left side of LEM with Astronaut Aldrin.

In this photograph, von Braun is carried on the shoulders of Huntsville city officials at an *Apollo 11* moon-landing victory celebration on July 24, 1969, in front of the Madison County Courthouse in Huntsville. American flags waved and celebrations continued in the town that had sent man to the moon—and had put *Explorer 1* in orbit with its Jupiter C just 11 years earlier. (MSFC.)

In this November 1971 photograph, from left to right, astronauts John Young, Eugene Cernan, Charles Duke, Fred Haise, Anthony England, Charles Fullerton, and Donald Peterson await deployment tests of the Lunar Roving Vehicle (LRV) qualification test unit in building 4649 at the MSFC. The LRV, developed under the direction of the MSFC, was designed to allow Apollo astronauts a greater range of mobility on the lunar surface during the last three lunar exploration missions, Apollo 15, Apollo 16, and Apollo 17. The up-rated engines of the Saturn V permitted it to boost the additional LRV weight, capitalizing on the engine designers' built-in margin in the engine's specifications during engine development. Improving the turbo pump's flow rates and adjusting the injector's mixing were among the contributing factors in the Saturn V's overall performance increase. (MSFC.)

This is a photograph of the launch of the SA-513, a modified unmanned two-stage Saturn V vehicle for the Skylab 1 mission, which placed the Skylab cluster into the Earth's orbit on May 14, 1973. Consisting of the S-IC first stage and the S-II second stage, the Saturn V booster's payload was the Orbital Workshop, the Airlock Module, the Multiple Docking Adapter, the Apollo Telescope Mount, and an Instrument Unit. (MSFC.)

This is a photograph of the Saturn IB vehicle that lifted off on May 25, 1973, carrying the crew of the Skylab 2 (SL-2) mission. The SL-2 mission launched the first crew to the Skylab: astronauts Charles "Pete" Conrad, Joseph Kerwin, and Paul Weitz. This crew made urgent repair work on the damaged Skylab to make it operational and habitable. The duration of this mission was 28 days. (MSFC.)

Skylab, shown here, became the free world's first space station. It was occupied in succession by three teams of three crewmembers. These crews spent 28, 59, and 84 days respectively, orbiting the Earth and performing nearly 300 experiments, enriching scientific knowledge of the Earth, the sun, the stars, cosmic space, and the effects of weightlessness. This view of Skylab in orbit was taken by the Skylab 4 (the last Skylab mission) crew. During the launch of Skylab, a protective micrometeoroid and heat shield was torn loose and one of the two solar power arrays was ripped off. The remaining solar array was only partially deployed. The astronaut crew was able to deploy a sunshade and free the power panel as shown in this photograph. The Earth atmosphere's drag on Skylab led to its fiery reentry and breakup in an area covering portions of the Indian Ocean and Western Australia in July 1979. (MSFC.)

Pictured here is astronaut Ed Gibson (left, in white suit) with six divers (dark suits) in MSFC's Neutral Buoyancy Simulator (NBS) during Skylab Extra Vehicular Activity (EVA) training. This overall view shows a mock-up of the Apollo Telescope Mount (ATM) Transfer Work Station. It was here that astronauts and technicians developed space walks that saved Skylab when it lost its sunshade during launch in 1973. (MSFC.)

The space shuttle orbiter *Enterprise* is shown arriving at MSFC in 1978 for the Mated Vertical Ground Vibration Test (MVGVT) series, the critical evaluation of the entire space shuttle component. In this view looking northwest over MSFC, *Enterprise* is seen heading south on Rideout Road near the Redstone Arsenal fire station as it is being transported to MSFC's building 4755 for later MVGVT tests. (MSFC.)

This photograph is an aerial view of the space shuttle orbiter *Enterprise* being hoisted into MSFC's Dynamic Test Stand for the Mated Vertical Ground Vibration test (MVGVT). The test marked the first time that the entire space shuttle elements, an orbiter, an external tank (ET), and two solid rocket boosters (SRB) were mated together. *Enterprise* was hoisted into the modified Dynamic Test Stand originally built for Saturn V testing, mated first to an external tank and subjected to vibration frequencies comparable to those expected during launch and ascent. Several months later, the solid rocket boosters were added for tests of the entire space shuttle assembly. The test series confirmed the structural interfaces and mating of the entire space shuttle system and allowed mathematical models used to predict the space shuttle's response to vibrations in flight to be adjusted. MSFC managed and conducted this important test program during 1978 and 1979 with support from the space shuttle contractors. (MSFC.)

Space shuttle *Columbia* carried MSFC-managed Spacelab 1 to orbit in 1983 for its maiden flight. Shown inside the science module during the STS-9 mission are, from left to right, mission specialist Robert Parker, payload specialist Byron Lichtenberg, mission specialist Owen Garriott, and payload specialist Ulf Merbold. The goal of the mission was to verify the Spacelab capabilities and to obtain valuable scientific, applications, and technology data from a joint U.S./European payload. (MSFC.)

This space shuttle *Atlantis* (STS-45) onboard photograph of its open cargo bay shows the forward portion of the ATLAS-1 (Atmospheric Laboratory for Applications and Science) payload at night. During the ATLAS mission series, science teams combined their expertise to seek answers to complex questions about the atmospheric and solar conditions that sustain life on Earth, specifically how Earth's atmosphere is affected by the sun and the Earth. (MSFC.)

This Boeing photograph shows Node 1, Unity module (at right), and the U.S. Laboratory module, Destiny, for the International Space Station (ISS) being manufactured in the High Bay Clean Room of the Space Station Manufacturing Facility at MSFC. The Unity module was launched aboard the space shuttle *Endeavour* (STS-88) on December 4, 1998. The U.S. Laboratory was launched aboard the space shuttle *Atlantis* (STS-98) on February 7, 2001. (MSFC.)

Underwater training is being conducted in MSFC's Neutral Buoyancy Simulator (NBS) on August 15, 1992, in preparation for on-orbit Hubble Space Telescope operations. The Hubble is the only telescope designed to be serviced in space by astronauts. It was carried into orbit by space shuttle *Discovery* in April 1990 and is named after the American astronomer Edwin Hubble. (MSFC.)

Six

RESEARCH PARK
AND SPACE CENTER

When von Braun's rocket team arrived in Huntsville in 1950, there was no local engineering and manufacturing support for their efforts. The chamber of commerce found Marietta Tool and Engineering of Marietta, Georgia, who agreed to relocate their operation to Huntsville. Opening in 1953, and shortly becoming Alabama Engineering and Tool Company, seen in this 1954 photograph, the company provided specialized hardware for emerging missile programs. (Ray Watson/TBE.)

With the merger of R. P. Brown's Indianapolis operations and the Alabama Engineering and Tool Company in 1956, the firm was renamed Brown Engineering Company (BECO). With John Hatch as general manager and George Epps as chief engineer, the challenge was the development of the Jupiter missile for ABMA. Growing rapidly, in 1958 BECO moved into the Huntsville Industrial Center the same year Milton Cummings was named president. (Ray Watson/TBE.)

When rocket development started at Redstone Arsenal, Huntsville had no high-technology infrastructure for support. Formerly an agricultural and cotton-mill town, the Lincoln Mills, a vast property bought by the Huntsville Industrial Associates after the textile mill closed in 1957, was converted into the Huntsville Industrial Center (HIC building), shown in this 1959 photograph, to provide space to jointly house Boeing, Chrysler, Spaco Manufacturing, BECO, and other firms. (Ray Watson/TBE.)

A highly successful local businessman and a major stockholder and president of BECO, Milton Cummings (left) placed the company on a sound business footing. In enlarged facilities and with funding he had arranged, a full range of engineering and integration capabilities was established. Shown here with von Braun at BECO, his company provided 20 million man-hours in a wide variety of efforts supporting the moon program. (Ray Watson/TBE.)

Cummings hired Joseph Moquin in 1959 as BECO executive vice president. Moquin, shown here, recognized that facilities at the HIC building, which used to be a cotton mill, were inadequate for the company to compete with some of the top engineering scientific and technical people throughout the country or to play a comprehensive support contractor role in the emerging lunar landing program at the newly formed MSFC. (Ray Watson/TBE.)

Located in downtown Huntsville, the temporary quarters in the Huntsville Industrial Center (HIC) building were not only shared among numerous contractors, but also with MSFC personnel as that new organization grew. This photograph shows drafting specialists from the Propulsion and Vehicle Engineering Laboratory at work in the HIC building in a 1964 photograph. (MSFC.)

BECO's Cummings and Moquin were interested in developing a major research and development facility in Huntsville. Under their leadership, BECO entered into a joint venture with the University of Alabama Huntsville Foundation to create Huntsville Research Park. BECO built the first facility there in 1962, shown in this 1963 photograph. Lockheed Martin established its operations there in 1963. (Ray Watson/TBE.)

In 1972, an entrance marker was erected at the corner of Governors and Sparkman Drives and the Huntsville Research Park was dedicated. It was renamed in memory of Cummings. A half century after BECO first constructed a building in the park's open fields, Huntsville's modern and well-equipped research park has evolved to become the second largest park in the country today and the fourth largest park of its kind in the world. (Ray Watson/TBE.)

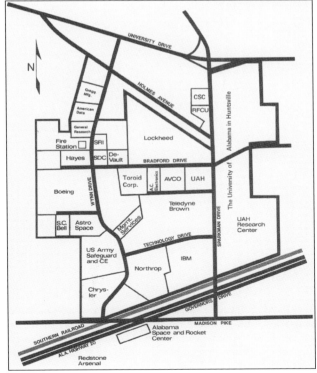

In the first 10 years of Cummings Research Park's existence, 24 companies constructed or leased buildings, as shown in this 1974 research park map. Today it houses the University of Alabama in Huntsville as well as major research and development operations for ADTRAN, Boeing, Northrop Grumman, Lockheed Martin, Raytheon, SAIC, Teledyne Brown Engineering, and the Hudson Alpha Institute for Biotechnology—in all, 285 companies performing research in 40 different fields and employing 25,000 workers. (Huntsville Public Library.)

Von Braun lobbied the Alabama legislature in 1961 for research support that would accommodate not only a research park but also a research institute and a university. With lawmaker's support and investment, the groundbreaking ceremony for the University of Alabama Research Institute was held in December 1962. Participating in the ceremony are, from left to right, Maj. Gen. Francis McMorrow, von Braun, unidentified, and Alabama governor John Patterson. (MSFC.)

Shown here is the University of Alabama Research Institute's new building, which opened in 1964. Established in north Alabama as an extension of the University of Alabama, the research institute provided the means to support faculty engaged in research activities by acquiring contracts and grants suitable for academic research. (UAH.)

During the 1960s and 1970s, the University of Alabama Huntsville Foundation and private partners continued to expand Cummings Research Park's boundaries and lobbied for a branch of the University of Alabama to be permanently located at the park and to support Redstone Arsenal and MSFC. The University of Alabama in Huntsville (UAH) campus, shown here, was established at Cummings Research Park in 1969 as an independent, autonomous campus. (UAH.)

Huntsville's facilities for public display of early Redstone Arsenal rocketry were somewhat limited. Since the fairgrounds were inadequate and there were no museums, missilery was displayed downtown on the square—previously the site of horse-drawn wagons of cotton for sale on cotton row. Shown on display in this 1953 photograph is a V-2 with poster entitled "V-2 Guided Missile" and providing a description and technical specifications. (AMCOM.)

The first Hermes guided missile was put on display at the courthouse square in downtown Huntsville in May 1953 as part of the city's celebration of Armed Forces Day. Soldiers from Redstone Arsenal as well as the installation's band participated in the Armed Forces Day parade. The sign reads: "First Army Ordnance Guided Missile Ever Placed on Public Display HERMES RESEARCH TEST MISSILE REDSTONE ARSENAL—HUNTSVILLE ALABAMA." (AMCOM.)

Shown in this July 31, 1968, groundbreaking ceremony photograph for the Alabama Space and Rocket Center are, from left to right, Edward Buckbee, space center director; Jack Giles, Alabama state senator of Huntsville; von Braun; Martin Darity, head of the Alabama Publicity Bureau; James Allen, chairman of the Alabama Space Science Exhibit Commission; Maj. Gen. Charles Eifler, commanding general of AOMC; and Huntsville mayor Glenn Hearn. (AMCOM.)

The Alabama Space and Rocket Center, shown here, opened in 1970 on land donated from the U.S. Army's Redstone Arsenal. An idea first proposed by von Braun, it houses interactive science exhibits, numerous permanent rocketry, and space exploration artifacts, as well as many rotating rocketry and space-related exhibits. Later named the U.S. Space and Rocket Center, it is America's largest missile and space exhibit. (AMCOM.)

The U.S. Air Force's Blackbird spy plane, shown on display at the U.S. Space and Rocket Center museum, remained the world's fastest and highest-flying operational manned aircraft throughout its career. The Lockheed SR-71 was an advanced, long-range, Mach 3 strategic reconnaissance aircraft developed by the Lockheed Skunk Works. In service from 1964 to 1998, the SR-71 was also called the Habu by its crews, in reference to a snake. (USS and RC.)

Baker, shown here, a South American squirrel monkey furnished by the navy, and Able, an American-born rhesus monkey supplied by the army, rode into space in May 1959. Able died in June 1959 from the effects of anesthesia given to allow the removal of electrodes implanted for the monkey's historic space flight. "Miss Baker," Queen of Space, spent her retirement years at the U.S. Space and Rocket Center. (USS and RC.)

Von Braun is shown posing next to his renowned rocket's first stage—the Saturn V Rocket that propelled men to the moon. The Saturn V is on display at the U.S. Space and Rocket Center museum in Huntsville. Von Braun (1912–1977) was one of the most important rocket developers and champions of space exploration during the period between the 1930s and the 1970s. (MSFC.)

Shown is the Huntsville family of vehicles, some of the rockets on display at the U.S. Space and Rocket Center museum's Rocket Park in Huntsville. These progressively larger vehicles were used in the stepped approach to building the mammoth Saturn V vehicle, which put men on the moon. Pictured are, from left to right, Mercury-Redstone, Jupiter IRBM, Saturn 1B, Juno II, Redstone, and Jupiter-C. (USS and RC.)

Pathfinder, a space shuttle mock-up, is shown in the shuttle park at the U.S. Space and Rocket Center. The integration of air and space technologies is no more apparent than in this rocket plane, which gets to space like a rocket and returns like a plane. The evolution of air and space in Huntsville, Alabama, is celebrated in the center's museum, space camp, aviation challenge, and Davidson Center. (USS and RC.)

In 1999, a full-scale model of the Saturn V rocket was erected at the U.S. Space and Rocket Center. The newest addition to the center is the Davidson Center for Space Exploration, shown here, which opened on January 31, 2008. The Davidson Center was designed to house the authentic Saturn V rocket—listed on the National Register of Historic Places—and many other space exploration exhibits. (USS and RC.)

The Davidson Center is like no other in the country. In its 476-foot-long, 90-foot-wide, 63-foot-high structure, suspended 10 feet above the floor, is a national historic treasure, the mighty Saturn V, restored to its Apollo-era readiness. The vehicle, shown here, is elevated above the floor surface with separated stages and engines exposed, so visitors have the opportunity to walk underneath the rocket. (USS and RC.)

Seven

PIONEERS

For his achievements as a pioneer in early aviation, William Quick was inducted into the Alabama Aviation Hall of Fame in Birmingham in 1982. Lorraine Q. Wicks, his granddaughter, accepted for the Quick family. He designed and built the first airplane to be flown in the state of Alabama in 1908. He produced a family of early aviationists involved in barnstorming and crop-dusting. (Southern Museum of Flight.)

Terah Maroney, a member of the Early Birds of Aviation, Inc., made his first flight in 1911. His biography is among the Smithsonian's Flying Pioneers Collection. Brother-in-law of William Quick, he was an able assistant in the design of the Quick monoplane. Barnstormer, exhibitionist, and builder, he took William Boeing for his first plane ride in 1914. He was killed in 1929 in east St. Louis while prop-starting an airplane. (SDAM.)

Col. Carroll Hudson was the first commanding officer of the Redstone Ordnance Plant, re-designated Redstone Arsenal in 1943. He guided the installation through a facility construction program for the development of ordnance rockets and guided missiles, while maintaining the capability for production of conventional ammunition. In addition to the Legion of Merit and other awards, he was inducted into the U.S. Army Ordnance Hall of Fame in 1993. (AMCOM.)

Lt. William Spragins II was killed in action over Germany on January 21, 1945. He was a B-17 pilot in the U.S. Army Air Corp. Listed in the Military Heritage Commission's Madison County Military Hall of Heroes, William is honored herein for the ultimate sacrifice in service for his country and as a representative of all from the Huntsville area who served in the U.S. Armed Forces during World War II. (Huntsville Public Library.)

BILL SPRAGINS II

See Chapter XX.

Maj. Gen. John Medaris was the first commanding general of the U.S. Army Ballistic Missile Agency (ABMA) at Redstone Arsenal. Medaris activated ABMA in 1956 and assumed responsibility for development of the Jupiter and Redstone. Under his leadership, the army successfully launched the *Explorer 1* satellite. He was honored by the National Space Club and the Smithsonian Institution for his leadership in the early days of the space program. (AMCOM.)

As commander of Redstone Arsenal beginning in 1954, Major General Toftoy laid the foundation of today's huge complex, becoming responsible for the entire army family of missiles and rockets. A plaque placed in Big Spring Park in Huntsville by grateful citizens honors the man known locally as "Mr. Missile," who had a great deal to do with turning the small cotton town into the rocket capital of the world. (AMCOM.)

Responsible for the launch of the western world's first satellites and manned space vehicles, von Braun led the rocket team that accomplished the launch of manned landings on the moon. Recipient of numerous tributes for his pioneering role in rocketry and space exploration, he maintained his Air Transport Pilot rating until retirement and is shown taking a supersonic flight in a T-38 at Edwards Air Force Base, California. (MSFC.)

Dr. Ernst Stuhlinger was on the cutting edge of science all of his life. Among the team who came to the United States with von Braun, his scientific work at Redstone and MSFC represents a broad spectrum, including developments in rocketry, Spacelab, and the Hubble telescope, and advocacy of electric propulsion systems for future space vehicles. He became a naturalized citizen in 1955 and happily made Huntsville his home. (Southern Museum of Flight.)

In 1941, U.S. Rep. John Sparkman was instrumental in the U.S. Army locating both the Huntsville Arsenal and the Redstone Ordnance Plant adjacent to Huntsville. Elected to the U.S. Senate in 1948, Sparkman was influential in the army consolidating its new missile and rocket research efforts at Redstone Arsenal. In gratitude, the army named its new administrative office complex on Redstone Arsenal the John J. Sparkman Center. (AMCOM.)

James Record served Madison County as an Alabama senator and later as the chairman of the Madison County Commission during the boom years of the 1960s through the 1980s. As a visionary and public servant, he brought positive change to Madison County, serving in fund-raising and promoting the building of the U.S. Space and Rocket Center and the Huntsville-Madison County Airport. (Huntsville Public Library.)

Businessmen and civic leader Carl Jones was a native Huntsvillian and a veteran of World War II. A farmer as well as senior partner of G. W. Jones and Sons Consulting Engineers, he served as city engineer, and in 1965, he received the city's Distinguished Citizen award. In 1967, the Huntsville-Madison County Jetport was named Carl T. Jones Field. All agree that no one deserves the title "Mister Huntsville" more. (Ray Jones.)

Ed Mitchell Jr. served as one of the Huntsville Airport's first board of directors (1956–1968) and was named the airport authority's first executive director in 1970. He was successful in its transition to the Huntsville Jetport and further expansion into the Port of Huntsville. The person who has done more for aviation in Huntsville than any other, the International Intermodal Center was named in his honor in 2007. (Huntsville Airport Authority.)

Often called the "Number One Citizen of Huntsville" and a symbol of the new South, Milton Cummings had an unmatched influence on the early growth of the city of Huntsville into the "Rocket Capital of the World." His success is attributed to the formation of the Brown Engineering Company and the development of the research park named in his honor to accommodate the expanding U.S. Army and Space Programs. (Huntsville Public Library.)

George Epps was born into a pioneer aviation family in Georgia, entering the U.S. Navy V-5 program in 1944 and later employed at Douglas and Lockheed Aircraft. In 1953, as chief engineer and later vice president of Brown Engineering Company, he supported NASA and army programs. An instrument rated pilot and collector of antique aircraft, he established several aviation firms and is an Alabama Aviation Hall of Fame inductee. (Southern Museum of Flight.)

A Huntsvillian, Lt. Michael Christian, was a prisoner of war in Hanoi, North Vietnam, from his capture in 1967, when his Grumman A6 Intruder was shot down over Ha Bac Province, until he was among 591 Americans released in 1973. Michael is honored herein for his patriotic service to his country, and he is honored as a representative of all from the Huntsville area who served in the U.S. Armed Forces during the Vietnam War. (Huntsville Public Library.)

An astronaut and pilot who flew on three space shuttle missions as a mission specialist and payload commander, Dr. Jan Davis spent almost 700 hours in space during STS-47 in 1992, STS-60 in 1994, and STS-85 in 1997. Serving as director of NASA MSFC's Flight Projects office and Safety and Mission Assurance office before retiring in 2005, she is currently vice president and deputy general manager for Jacobs, supporting NASA MSFC. (Southern Museum of Flight.)

Capt. Trey Wilbourn, a native Huntsvillian, was shot down over Kuwait in 1991 in his AV-8B Harrier during a night attack in Operation Desert Storm. Listed in the Military Heritage Commission's Madison County Military Hall of Heroes, Trey is honored herein for his ultimate sacrifice in service to his nation and as a representative of all from the Huntsville area who served in the U.S. Armed Forces in the Persian Gulf. (Joyce Wilbourn.)

www.arcadiapublishing.com

Discover books about the town where you grew up, the cities where your friends and families live, the town where your parents met, or even that retirement spot you've been dreaming about. Our Web site provides history lovers with exclusive deals, advanced notification about new titles, e-mail alerts of author events, and much more.

MADE IN THE USA

Arcadia Publishing, the leading local history publisher in the United States, is committed to making history accessible and meaningful through publishing books that celebrate and preserve the heritage of America's people and places. Consistent with our mission to preserve history on a local level, this book was printed in South Carolina on American-made paper and manufactured entirely in the United States.

This book carries the accredited Forest Stewardship Council (FSC) label and is printed on 100 percent FSC-certified paper. Products carrying the FSC label are independently certified to assure consumers that they come from forests that are managed to meet the social, economic, and ecological needs of present and future generations.

FSC
Mixed Sources
Product group from well-managed forests and other controlled sources

Cert no. SW-COC-001530
www.fsc.org
© 1996 Forest Stewardship Council

Find Your Place in History.